God's Own Scientists

God's Own Scientists
Creationists in a Secular World

Christopher P. Toumey

Rutgers University Press
New Brunswick, New Jersey

Library of Congress Cataloging-in-Publication Data

Toumey, Christopher P., 1949–
God's own scientists : creationists in a secular world /
Christopher P. Toumey.
p. cm.
Includes bibliographical references and index.
ISBN 0-8135-2043-6 (cloth) — ISBN 0-8135-2044-4 (pbk.)
1. Creationism—United States. 2. Creationism—North Carolina.
3. United States—Church history—20th century. 4. North Carolina—
Church history. I. Title.
BS651.T59 1994

306.6'31765—dc20   93-24241
                                                   CIP

British Cataloging-in-Publication information available

To the memory of William Hubert Toumey,
1911–1969
Ni fheicfimid a leitheid aris

Contents

Contents

viii

Acknowledgments

"I have learned much from my teachers," said Rabbi Judah Ha-Nassi, "but more from my friends, and most of all from my students." So too with myself. I begin by thanking my teachers, particularly the professors of anthropology at the University of North Carolina who gave me the freedom to make my own mistakes in this research. Bless them for letting me loose. Special thanks to Carole L. Crumley: may every graduate student like me get an adviser like her.

Next, my fellow graduate students, especially Paul Benjamin and Allan Fischer; other friends were Suphronia Cheek, Kimi Julian, Ney Reiber, Sandy Shugart, and Anita Whitmire, who each helped this Irish Catholic appreciate southern Protestantism. Jane D. Brown kindly included questions about creationism in the Carolina Poll of October 1983. The staff of the Davis Library at UNC, especially the humanities reference librarians, were always wonderful; Glenda Hightower and Eugenie Frederick gave me generous hospitality. Lester Harrison, Ray Eve, Frank Harrold, and George Webb engaged me in many provocative discussions about creationism. And an anonymous referee twice combed through my manuscript and twice gave me numerous valuable suggestions for improving it.

Eileen McLeod deserves my gratitude for reasons she knows well; Eileen London has been the best mother an anthropologist could want. Kathryn J. Luchok is still my best friend, even though I went to a creationist meeting the night I was supposed to take her to a Talking Heads concert. Also to be acknowledged are Tom McIver, Edward Larson, and Ronald Numbers. In my

Acknowledgments

less-modest moments I please myself by thinking that my work stands shoulder to shoulder with theirs.

Most of all, my kind creationist friends in North Carolina, especially the folks in the study group I describe, deserve my gratitude for letting me look into their hearts. Some of what I've written will flatter them, and some of it will sting them, while other passages will leave them wondering what all the fuss was about. Because I guaranteed confidentiality, I do not cite them by name. Regardless, I've enjoyed my times with them, and I appreciate their help enormously.

Finally, my students at the University of North Carolina, the College of Wooster, and North Carolina State University always challenged me to meet the highest intellectual standards. Being their teacher has thrilled me, and it has also made me a better scholar.

Parts of this book have appeared in the *International Journal of Moral and Social Studies*; the *Journal of the American Academy of Religion*; *Social Studies of Science*; *Soundings: The Journal of the Elisha Mitchell Scientific Society* (North Carolina Academy of Science); and *Creation/Evolution*. I am grateful for their permission to use that material.

These institutions gave financial support for my research: the Graduate School at the University of North Carolina, which awarded me a Pogue fellowship; the University Research Council of UNC, which provided two grants to Norris B. Johnson and myself; Sigma Xi, The Scientific Research Society; and the College of Wooster, which bestowed upon me two faculty development grants. Their assistance has been very helpful to me, and I appreciate it deeply.

Abbreviations of Creationist Organizations

ASA   American Scientific Affiliation, an organization of evangelical Christians in the scientific community

BSA   Bible-Science Association, the Missouri Synod Lutheran creationist organization

CAL   Christian Action League, a conservative Christian political lobby in Raleigh, North Carolina

CCS   Center for Creation Studies, an academic unit of Rev. Jerry Falwell's Liberty University in Lynchburg, Virginia

CHC   Christian Heritage College, a conservative Christian school in El Cajon, California, founded by Rev. Tim LaHaye

CLL   Churches for Life and Liberty, a conservative Christian network in North Carolina

CRS   Creation Research Society

CSRC   Creation-Science Research Center, San Diego, directed by Nell Segraves

GRI   Geoscience Research Institute, the Seventh-day Adventists' creationist office at Loma Linda University

ICR   Institute for Creation Research, Santee, California, founded by Dr. Henry M. Morris

SOR   Students for Origins Research, a creationist group in Santa Barbara, California

Part One
Scientific Authority

One
Mister Fossil Comes to Raleigh

Mister Fossil is modern creationism made flesh. He blasts the idea of evolution as a lie vomited up by Satan to erode the moral foundations of our lives, and he summons up some symbols of scientific authority as corroboration for the creationist account of cosmic and human origins. Thus he embodies the two essential features of American creationism in the late twentieth century: a moral theory that evolution is intimately involved in immorality, and a philosophy of science that puts science on the side of creationism, not evolution.

Mister Fossil's real name is Arleton C. Murray, but he prefers to advertise his services under his sobriquet. He tells his audiences that he learned the science of preserving fossil bones at an institute in England, after which he worked as a fossil preparator at the Smithsonian Institution in Washington, D.C., in the 1930s. From this position deep within the realm of paleontology, he saw the evil idea of evolution close up, and he accepted it, at first. One night he went to an evangelical revival to scoff at the Christian religion, but as he heard its message of salvation, he turned to the truth and soon rejected evolution, with all that it stands for. When he declared his new-found faith to his evolutionist bosses at the Smithsonian, they demanded that he choose between evolution and Christ, whereupon he promptly and dramatically turned his back on his job with the paleontologists and turned his face toward Jesus. In this way was Mister Fossil born. From time to time he appears on conservative Christian speaking circuits, where his credentials in both the science of paleontology and the Christian faith

I

give him the authority to scorch the idea of evolution on moral grounds and to endorse creationism in scientific terms.

Mister Fossil came to Raleigh, North Carolina, in the fall of 1984 because a local creationist study group in the Research Triangle area of that state had heard that he was making an East Coast speaking tour, and they wanted to sponsor a talk to their community. This group (which I describe in more detail in chapter 11) usually meets as a small informal gathering in the home of one of its members, but occasionally it reaches out to a much larger audience by hiring a speaker, renting a hall, and advertising to the public.

This study group is very much devoted to the leadership of Henry M. Morris, head of the Institute for Creation Research, near San Diego, California. Morris, a hydraulic engineer with a Ph.D. from the University of Minnesota, is undoubtedly the most influential leader of modern U.S. creationism, by virtue of his dozens of books on the subject and his untiring personal devotion to the cause during the past five decades. Several members of the study group know him personally, and all of them treat his teachings with great respect. As I will argue later in this book, all modern creationist thought must be measured in relation to Morris's teachings, and the local group in North Carolina could reasonably have expected Mister Fossil to deliver a message in line with those teachings.

And so in October 1984, the creationist study group arranged to use the fellowship hall of a church in Raleigh and advertised a talk by Mister Fossil, which attracted an audience of about fifty people. It was an interesting coincidence that Mister Fossil looked much like Henry Morris. He was short and round, with receding grey hair combed straight back, and he wore thick glasses in wire frames. Mister Fossil wore a khaki bush jacket with a *Tyrannosaurus rex* embroidered on the left breast, a small badge reading "creation scientist" on the right lapel, and a Gideon's badge on the left lapel. (The Gideons are the association of evangelical Christian businessmen best known for distributing Bibles to the public.)

After the host from the study group offered an opening prayer and a vague introduction, Mister Fossil launched into his testimony. He had been interested in nature since childhood,

2

he told the gathering, but he was told that "it never had God behind it." While at the Smithsonian as a fossil preparator, "I didn't know about Jesus Christ as my savior. I used to annoy prayer meetings; I was interested in drinking liquor and smoking cigars." One night, he continued, he went to a religious revival in Greenbelt, Maryland, to scoff at the preacher. To his surprise, he was moved by what he heard, and he became a Christian that same evening. As a consequence, he soon began speaking out against evolution. After the Smithsonian evolutionists heard about an anti-evolution talk he had given at Washington Bible College, his boss confronted him and demanded that he cease such activities. Said Mister Fossil, "He said I had to, but I said I didn't have to, and I just walked out and never went back."

Mister Fossil illustrated his talk from that point on with exciting slides of finding dinosaur bones, excavating them, and reassembling them at the museum. As he related his fossil-hunting experiences in places from Nebraska to Panama and showed illustrations of the intricate work of preserving and reassembling the skeletons of great dinosaurs, he easily won the respect of his audience; here was a creationist who undoubtedly possessed the scientific expertise of those who unlock the past.

Next he launched into a scathing attack on the idea of evolution. "All museums teach the doctrine of evolution," he said. "They're all Karl Marx [*sic*]. . . . Fossils are the Waterloo of evolution." The geological column, explained Mister Fossil, is "from the perverted mind of an evolutionist." The fossil rooms of museums are "taboo to the public; they might find out something." "Evolution is a fake and a lie of the devil," he continued. "Evolutionists actually believe a watch would evolve from a hairspring and filings!"

This was fundamentalist preaching at its finest, following the classic form of confessing a sinful life, then describing the personal experience of conversion that changed everything, and finally atoning by exposing the lurid secrets of the evildoers. If Mister Fossil had stopped there, he would have given the local creationists everything they could have asked for. But he was not finished yet. After blasting evolution, he led the audience through his personal theory of creationism. Fossils, he said, are evidence of Noah's Flood. For example, the deposits at Agate

3

Springs, near Carnegie Hill, Nebraska, contain numerous species together: "Only a great catastrophe like a flood could mix them all together like this." The key to absolute proof of creation, according to Mister Fossil, is fossil fish, which were buried instantly and died of suffocation. Furthermore, the fossil forms of leaves, shrimp, tapirs, and starfish are just like today's forms, showing no sign of evolutionary change.

By this point, the members of the creationist study group were beaming, enjoying every minute of their lecturer's presentation. As he neared the end of his talk, however, Mister Fossil elaborated on his personal views. Dinosaurs and humans, he said, had never lived together. Dinosaurs did not become extinct in Noah's Flood, he explained; they perished in a previous catastrophe, "Lucifer's Flood." *After that*, the world became "without form, and void" (Gen. 1:2). The six-day creation of Genesis, according to the speaker, was consequently a *re*-creation of the earth and its creatures, which only then included humans.

This theory of two creations and two floods had no special significance for most of his audience, but the members of the study group were aghast. In terms of the creationist orthodoxy laid down by Henry Morris, Mister Fossil was speaking heresy. He was arguing for gap theory, the view that there was a long period of time between the events of the first and the third verses of Genesis, chapter 1. Such a gap would have allowed many geological events and changes to happen before the six-day creation. To Morris and his followers, including the study group, gap theory is absolutely impermissible on two counts. It undermines the geological significance of Noah's Flood, which to Morris was the event that accounts for all important geological evidence, and it admits the possibility of a time span long enough for evolution to have occurred. The men from the local group knew this well, and they winced when Mister Fossil expounded his gap theory.

Worse was to come. The speaker attacked Henry Morris personally: "Morris and his friends know nothing about fossils; Morris is an engineer, not a paleontologist." Morris, he explained with disdain, is "wrapped up in his view." Mister Fossil then justified his own theory of Lucifer's Flood by saying that it came to him "as the spirit moved me." Regardless of its scientific value or lack thereof, this was a perfectly credible explanation to those in the fundamentalist audience who were not particularly sensitive to the arcane nuances of creationist orthodoxy. In a church hall of conservative Christians, assembled to hear the good news of the creationist message, it would be folly to refute such a justification. It would be tantamount to refuting grace from above.

The evening ended on a flat note. After a brief question-and-answer session, the host, who was clearly rattled by Mister Fossil's heresy, blurted, "Well, uh, this shows you that not all creationists agree on everything." Mister Fossil had turned out to be an embarrassment. He had stung the members of the creationist study group, sharply and publicly, leaving them speechless in front of their friends and supporters.

I tell the story of Mister Fossil's visit to illustrate some core features of modern U.S. anti-evolutionism, including the subtle and complex nature of creationist thought. Two of the most common and simplistic reactions to creationism, especially from its enemies, are that creationism is nothing more than a rote exercise in biblical literalism, and that the source of creationism is ignorance of science.

In the first case, opponents believe that fundamentalists reject the idea of evolution because evolutionary time contradicts the six-day story in Genesis, or because the evolutionary account of human origins is a direct challenge to Gen. 1:27 ("God created man in His own image"). Of course it is true that the particulars of creationism have to be anchored in scriptural proof-texts, but I will argue that this body of knowledge and belief is much richer and much deeper than a narrow-minded devotion to a few dozen verses of sacred scripture. In fact, the issue of gap theory shows that people like Henry Morris and Mister Fossil can each stake a claim on their respective proof-texts and at the same time maintain mutually irreconcilable theories about a profound issue like creationist time.

The second basis of opposition, that creationism equals lack of scientific knowledge, is likewise misleading. A long series of polls and surveys confirms that "level of education" is one of the best predictors of belief about evolution and creationism. People whose formal education ended with high school are most likely to accept creationism, while those who believe

5

in evolution are most often college graduates and those with postgraduate educations. This reliable generalization is confounded by the fact that much of the leadership of the creationist movement, particularly its best-known speakers and writers, includes people with respectable Ph.D.'s from fine secular universities. If one were to judge creationism in terms of the credentials of Henry Morris or his colleague Duane Gish, as one might take into account the credentials of Stephen Jay Gould and Carl Sagan when assessing their arguments about evolution, then the obvious superiority of evolutionary thought is somewhat less obvious. The presumption that science is entirely compatible with evolution, and not at all with creationism, has been subjected to an aggressive and articulate campaign by creationists like Henry Morris, Duane Gish, and many others, to make the opposite argument. The creationist claim that the authority of modern science resides with creationism, not evolution, has convinced numerous Americans, including many who are not fundamentalist Christians. Evolutionists cannot take it for granted that nonscientists will associate scientific authority with evolution.

Because the problem of modern creationism is more complicated, more subtle, and more interesting than is ordinarily supposed, I propose to approach it by seeing creationism as a system of cultural meanings about both immorality and science that helps fundamentalist Christians make sense of the realities, anxieties, changes, and uncertainties of life in the United States in the late twentieth century. As a system of meanings about immorality, creationism offers a series of theories that allege that the idea of evolution is intimately involved, as cause or consequence or both, in the moral disintegration of modern U.S. life. These feelings about immorality are the common stock of fundamentalist thought; creationism, then, is the subcategory of fundamentalism that ties those feelings to the problem of evolution.

Equally important, and no less problematic, is a series of meanings about the moral authority of science. Creationists and other fundamentalist Christians speak of the "plenary authority" of the Holy Bible, which is a way of saying that the written record of revelation embodies enough wisdom to answer *any* of life's questions, big or small, personal or general, religious

6

or secular. In the following chapters I will argue that creationism must also come to terms with a second kind of plenary authority, namely, that of science. The prestige of science in the U.S. culture of our time is such that science is also widely believed to embody enough wisdom to answer any of life's questions. This is not to say that science really deserves such awe, or that it could ever answer so many questions, or even that most scientists believe this about science. Rather, the problem for creationists and others is that too many people take science too seriously, endowing it with a moral authority equivalent to that of our conventional Judeo-Christian religions. Derived from that moral authority is scientific sanctification, a secular grace that is invoked to enhance the appeal of various ideologies, policies, and commodities. That being so, the credibility of creationism (like that of evolution, or psychiatry, or a brand of painkiller or laxative, or many other ideologies and commodities) depends on coming to terms with the plenary authority of science. One way is to cloak creationism in scientific sanctification by contending that science corroborates creationism, in general and in detail. Thus the terms *scientific creationism* and *creation-science*.

The contention that science is on the side of the Holy Bible in a conflict against evolutionary thought has been downright shocking to most U.S. scientists and science teachers. Strange enough, they feel, that militant anti-evolutionism should come back to life late in the twentieth century, but stranger still that it claims to have scientific authority as its ally. With dismay, the friends of evolution watch the creationists come forth with some scientific credentials, some select definitions of science, and a remarkably strong body of public opinion concurring with their claims to scientific status. Anti-evolutionism was never like this before. Always it bore a reputation, justified or not, of opposing science, of decrying science as too materialistic, too naturalistic, too skeptical. Thus did the eighteenth-century French religious authorities reproach the Comte de Buffon when he suggested that species were changing; so the nineteenth-century conservative English episcopate reviled Charles Darwin for documenting the mutability of species; likewise did the early twentieth-century U.S. fundamentalists rebuke John Thomas Scopes for defending

7

the suggestion that a modest exposure to evolution is a reasonable part of science education. Each time, scientists thought that they and their public could see a clear demarcation between scientific knowledge and religious belief. Until modern times.

The daring new creationist proposition poses some blunt questions about moral interpretations of science. It forces one to consider the ways conservative Protestants have revised their feelings about the plenary authority of science, and the reasons why they have done so. Somewhat more generally, it requires one to set aside the conviction that science embodies an objective truth that stands apart from the subjective textures of human life, or that the content of science is a durable reality untouched by moral considerations. Furthermore, there is a variety of creationist positions on science, so it might be said that modern creationism comprises a *series* of meanings about the moral authority of science.

In summary, I argue that, in order to appreciate why creationism moves people as deeply as it does, one must see it as a body of existential questions and answers—"cultural systems of meaning," in the language of anthropology—about realities, anxieties, uncertainties, and changes in U.S. life in our time. Most particularly, creationism asks and offers answers to these three questions: (1) How is the idea of evolution involved in abortion, communism, sex education, the disintegration of the family, and a host of other matters that fundamentalists consider immoral? (2) How can the plenary moral authority of science be employed to substantiate the historical authenticity of creationist accounts of origins, as revealed in Genesis? and, the ultimate question for creationism, (3) How can that scientific sanctification be employed to corroborate creationism's diagnosis of immorality?

The troubles that concern creationists may be timeless in the most general sense, but in fact they take the form and face of a particular period of history. Abortion was a far less salient matter before the *Roe v. Wade* decision of 1973; disputes about prayer in the public schools were almost unheard of before 1962; the political confidence of fundamentalists regarding the issues of abortion, school prayer, and creationism relied on the

8

1980 election and 1984 reelection of President Reagan, widely interpreted as ratifications of fundamentalist values. Considering that creationism is a set of existential questions and answers about such troubles, then it, too, is molded to the shapes of modern history. The creationism of the 1980s differs from that of the 1920s in terms of both issues of immorality and the plenary authority of science. My concern is primarily to understand the form of creationism that arose in 1961 when John C. Whitcomb, Jr., and Henry M. Morris, in their book *The Genesis Flood*, launched the claim that scientific authority corroborates the stories of Genesis. The high tide of influence for this kind of creationism was the year 1981, by virtue of the "balanced treatment" laws enacted in the legislatures of Arkansas and Louisiana; its unhappiest time was 1987, when the U.S. Supreme Court struck down such laws. Although creationist thought has survived the 1987 ruling, I choose this segment of time because that was the season when creationism was most successfully projected beyond its sectarian base, and into the wider U.S. culture.

Because of my choice to situate creationism in such recent times, I give the Scopes Trial of 1925 and the conditions surrounding it much less attention than more contemporary events. To the reader who wants a comprehensive history of U.S. creationism, I commend Edward Larson's excellent social and legal history, *Trial and Error* (1985), and Ronald Numbers's definitive work, *The Creationists* (1992).

Another specification is this: There are numerous varieties of creationist thought, on both matters of immorality and scientific authority, and numerous kinds of creationist groups. The most thorough exploration of creationist thought is the dissertation of Thomas McIver (1989), which parses it inside and out. I, however, will present only the main variations, with the intention of showing why certain groups and organizations embrace their respective kinds of creationist thought.

For readers curious about my intellectual preferences, they are the interpretive anthropology of Clifford Geertz and the sociology of knowledge of Karl Mannheim, which I combine into an anthropology of science. The former asks us to comprehend the systems of meanings and symbols by which people *un*like ourselves make existential sense of their lives, and then

9

to interpret those meanings and symbols to others (Geertz 1973, 1983)—thus my definition of creationism as a body of existential questions and answers about immorality and science.

Mannheim's sociology of knowledge adds depth to Geertzean anthropology by positing an intimate connection between a given form of knowledge or belief and the kinds of people who embrace it. Its basic principles are these:

1. The content of a body of thought is existentially determined, which is to say that it is based on human experiences that are particular to time and place. One must thus specify the historical situation within which we find that form of thought.
2. Within a given historical situation, there are multiple forms of existentially determined thought because there are multiple social groups, each having its own unique experience. So, within a given historical situation, the sociology of knowledge must also correlate kinds of thought with kinds of people.
3. It is necessary to account for a body of thought by demonstrating its existential value—questions and answers about realities, anxieties, and so on—for the people who accept it (Mannheim 1936, 1952, 1971a, b).

That last principle dovetails exactly with Geertzean anthropology; the first two insist that knowledge and belief must be situated in history and located in society. To humanize the creationists, I consider them the same way other anthropologists behold the many fascinating peoples who capture our professional attention. This is not to agree with creationists any more than my brother and sister anthropologists place their faith in cargo cults, ghost dance movements, or African theories of witchcraft, but it is nevertheless an attempt to understand the existential meanings about immorality and science that make creationism what it is, and to interpret them fairly. Consequently, the spirit of this work is not to denounce the creationists' claims for having departed from scientific norms, but to see what their beliefs mean in the context of their lives.

Mister Fossil Comes to Raleigh

Ronald L. Numbers recommends something similar to the readers of his history of creationism:

> Academics who would have no trouble emphatically studying fifteenth-century astrology, seventeenth-century alchemy, or nineteenth-century phrenology seem to lose their nerve when they approach twentieth-century creationism and its fundamentalist proponents. ... Although many scholars seem to have no trouble respecting the unconventional beliefs and behaviors of peoples chronologically or geographically removed from us, they substitute condemnation for comprehension when scrutinizing their own neighbors. (Numbers 1992:xvi–xvii)

My work, like his, is meant to contribute to comprehension, not to condemnation.

I divide my treatment of creationism into two parts. There is creationism as a national movement, wherein formal statements of belief are generated by a sophisticated leadership. There is also creationism as a local experience, more intimately bound to time and space, and sensitive to parochial circumstances. To stitch the two together, I suggest that the national leadership of the creationist movement supplies a matrix of creationist thought; the local creationists then choose certain portions of that thought and try to fit them into the particular realities of science, religion, job, education, and family that shape their own lives. The experience of creationism in North Carolina in the 1980s is my example of the latter.

Thus I present creationism in three parts. Part 1 (this chapter and the next) establishes the idea of the plenary authority of science; that is, the idea that something is more valuable and more credible when it is believed that science endorses it. This puts creationism into U.S. history, specifically the cultural history of the moral authority of science. There was a time when science was intimately tied to evangelical Protestant values, but the connection was severed when science acquired an autonomous moral authority, and modern creationism attempts to reestablish the former relationship. U.S. culture thus has three historical models for understanding how science is related to morality; I call them the Protestant, secular, and trivial models. Part 2 concerns the modern creationist movement. It describes a map of moral reasoning by which fundamentalists and evangelicals steer their way through the social landscape of the United States, and it explains how the specifics of creationist thought on immorality and science are generated from relations and conflicts among the main creationist organizations. The result is a grand outline of the matrix of creationist thought and its existential value to the various kinds of creationists.

Part 3 places creationist ideas in the local circumstances of North Carolina in the 1980s. It describes the historical conditions of North Carolina religion and politics that surround the creation-evolution controversy, wherein creationist thought is a package of doctrines dutifully received from Henry Morris and the Institute for Creation Research in California. Doctrine, however, is shaped to the particulars of time and place, as in the attempt to project the values of creationism into the secular world by proposing that public school science education should include scientific creationism on equal terms with evolution.

Part 3 also shows how the local creationist study group in the Research Triangle area of North Carolina arrives at its interpretations of creationism, and how scientists and engineers do the same. There is a summary of the ideological outlines of creationist knowledge and belief in North Carolina in the 1980s, and a final consideration of the existential nexus of modern U.S. creationist thought: Can creationism explain where immorality comes from? Can scientific sanctification be employed to convince doubters that the stories in Genesis are historically authentic? And can this plenary authority of science be united with that moral theory?

A few critical definitions: the term *fundamentalist* should be used with caution for two reasons. First, those who are called by this word seldom use it themselves. They usually describe themselves as "Bible-believing Christians." Secondly, those who use it of others often invest it with awful implications; "fundamentalist" has become a nasty thing to call someone. I usually prefer *conservative Christian*, a less pejorative term suggesting that social conservatism is closely connected to the theological conservatism that makes these people distinct. Sometimes I use *fundamentalist* to avoid excessively repeating *conservative Christian*. I do so without intending to denounce people through that term.

Biblical creationism is a belief that Genesis is an accurate and straightforward record of origins; *scientific creationism* is a belief that the events mentioned in Genesis can be substantiated with technical evidence, without direct mention of scripture. Regardless of whether the distinction is plausible, it is central to modern creationist thought, which commits itself to the idea that the technical evidence can be sufficiently abstracted from its scriptural roots to justify a prominent place in secular U.S. science, and in public school science education. I am more concerned here about why it is thought that God's creation can be removed from the Holy Bible to be packaged in a scientific image, than with the sizzling debate about whether such packaging is plausible. To the extent that the two kinds of creationism can be distinguished, this book is about scientific creationism; the biblical form plays a background role. I use the generic term *creationism* usually to mean scientific creationism, but sometimes to indicate a combination of scientific and biblical; I trust that my meaning is clear in each context.

One last caveat: my reader might expect a book about creationism to be a book about religion, decorated with sectarian schisms, scriptural stanzas, or moral majorities. Indeed, there is plenty of that herein. More than that, however, this is a book about science, in the sense that certain citizens called creationists are making room for the secular authority of science in their moral sentiments. Think of the scientific creationists not as slaves to the scriptures but, for the moment, as God's own scientists.

Two
Creationism and Scientific Authority

What is meant by *science* is not as straightforward as it seems, for different people have different understandings of this term, and some understandings are more influential than others at certain points in U.S. history. To see creationism in relation to science, it is worth seeing how a series of three meanings of science have unfolded. I call these the Protestant model of science, the secular model, and the trivial model.

Early in the nineteenth century, evangelical Protestantism and science were so intellectually compatible in the United States that a naturalist and a minister could easily agree on what they believed about nature. Indeed, the naturalist often was a minister, for the heart of the Protestant philosophy of science was the idea of "two revelations." God had revealed himself to us twice, in scripture and in nature, so that curiosity about nature was a good Christian virtue, provided that it was guided by the same kind of piety that steered one's interest in scripture. Although the methods of studying nature were different from those for scripture, the end result was expected to be the same: a person was morally enriched by seeing the evidence of God's character in his creation. The adaptation of a creature to its environment, for example, was a sign of God's careful design, while abundance in nature was a clue to the generosity of the Creator.

Coexisting with the Protestant philosophy was the secular model of the European Enlightenment. This second approach assumed that human reason gave a person the confidence—or the arrogance—to understand the natural world in terms of

14

natural laws, which could be easily discovered independent of religious guidance. Instead of revelation or salvation, its moral goals were progress and reason. The secular model, however, had few adherents in this country in the early nineteenth century. Thomas Jefferson and a handful of other intellectuals championed its cause, but their philosophical influence was negligible. Almost all of U.S. higher education was in the hands of Protestant denominations, which of course nurtured the Protestant way of thinking.

A third American way of thinking about science and nature depended on neither Protestant nor Enlightenment inspirations. This last model was the gospel of material prosperity, which looked to science only for useful knowledge that would unlock the natural resources of a great continent and convert them into comfort or wealth. Science meant nothing more than engineering and technology. Moral reasoning and philosophical reflection were superfluous. Science was to be measured and appreciated in terms of its tangible results. The importance of this attitude is deceptive, for it is intellectually shallow, if not absolutely trivial. Nevertheless, it was—and is—pervasive. Here science served the role of endorsing and increasing prosperity.

If the study of nature led to different conclusions than did the study of scripture—that is, if science digressed from Protestantism—then the Protestant model had to concede some authority to the secular model. The former uncovered one kind of truth, the latter another. But if truth was a simple, unified whole, then digressions, contradictory conclusions, and multiple kinds of truth were unthinkable. For this reason, the Protestant model was greatly strengthened and elaborated by three philosophical stances that each defined truth as a simple unified whole. They were the Scottish Common Sense philosophy, Baconian empiricism, and the Princeton Theology.

The Common Sense philosophy arose in Scottish universities in the eighteenth century as a reaction against the specialization of knowledge. At that time many English and Continental intellectuals accepted that knowledge was rightly divided into numerous professional spheres, because truth had many facets, most of which were difficult to apprehend; also, that only a specialist could apprehend a given facet of the truth, so that nonspecialists required the mediation of specialists. The Common Sense philosophy cut through that epistemology like a hot knife through butter. It proposed that the things worth understanding were not particularly opaque. Rather, they were just what they appeared to be, and a person of average intelligence could clearly see them as such. Sense perceptions were reliable, and reliable sense perceptions were common. Thus, the Common Sense philosophy (Grave 1960:112–114; McIver 1989:15–19; Olson 1975:chap. 1; Webb 1983, 1986).

This philosophy, the most influential school of thought in the United States in the mid– and late nineteenth century (Grave 1960:4; Webb 1983:34), implied that truth was a single, uncomplicated whole (Webb 1983, 1986). The burden of proof rested with those who argued that truth was complicated and that one's perceptions had to be specialized. The genius of this framework was that specialists had the impossible burden of convincing nonspecialists that only specialists could apprehend the truth. Even if, for argument's sake, the specialists' contention was true, nonspecialists still could not accept it, because, by definition, they could not understand it. As a result, the logic of the Common Sense philosophy automatically discredited intellectuals who expected scientific knowledge to be specialized, including those who felt that knowledge derived from nature might be different from that derived from scripture.

Closely related to Common Sense was Baconian empiricism, which began with the view that the truth is uncomplicated and self-evident, which then meant that theories, hypotheses, metaphysical thoughts, and other mental complications were unnecessary (Hughes 1983:112; McIver 1989:15–25; Webb 1983, 1986). According to the dichotomy of values in Baconian empiricism, facts and the activity of collecting facts were good; theories, and time spent on theorizing, were bad: "In the American vernacular, 'theory' often means 'imperfect fact'—part of a hierarchy of confidence running downhill from fact to theory to hypothesis to guess" (Gould 1981:34).

Science, then, was a simple business of observing, collecting, and classifying the facts of nature (Webb 1983:34).

Finally, the Princeton Theology applied the Common Sense philosophy and Baconian empiricism to the appreciation of

16

scripture. In accordance with Baconian principles, it asserted that scripture, like nature, comprised a body of uncomplicated facts (Marsden 1980:111–112; McIver 1989:12; Sandeen 1967:73, 1970:169); in the spirit of Common Sense, it assumed that "even simple Christians could understand the essential message of the Bible on their own" (Marsden 1980:110). The result was that, for the first half of the nineteenth century, evangelical Protestant thinking about scripture was practically identical to American scientific thinking about nature. And when a fact of nature conformed to a fact of scripture (as when the adaptation of a creature to its environment illustrated God's careful design), so much the better for nature *and* scripture, since this was exactly what the idea of the two revelations had predicted.

It is not surprising, then, that when the study of nature was guided by Protestant principles, it affirmed the same lessons of cosmic order and design as did scripture; or that both kinds of learning inspired the same evangelical style of proselytizing; or that they evoked the same personal ethos of selflessness and spiritual dedication (Burnham 1987:23, 144–146; Rosenberg 1966:150). Furthermore, before the 1840s, highly organized knowledge about nature was not called science, but rather natural history (for the life sciences), natural philosophy (the physical sciences), or natural theology (the view that nature was a second revelation), thereby indicating that knowledge about nature was a kind of knowledge of history, philosophy, or theology. Shortly after the word *science* acquired its modern meaning, an anonymous Presbyterian writer asserted that "the mosaic account of creation [is] scientific" (*Presbyterian Quarterly Review* 1858).

The three components of the Protestant model of science—the Common Sense philosophy, Baconian empiricism, and the Princeton Theology—were all reducing tendencies. They reduced knowledge to its most simple, obvious, tangible qualities. At the same time, they implied that sophisticated structures of knowledge (that is, theories and hypotheses) were less than factual, less than real, and unwanted complications that interfered with a person's apprehension of the truth. Charles Hodge, president of the Presbyterian seminary at Princeton and arch-apostle of the Princeton Theology, once dismissed a certain

17

critic as someone who had "a head made light by too much theorizing" (Marsden 1980:112).

During the middle third of the nineteenth century, many U.S. scientists turned toward the secular model of science. Their change of heart was not a conscious rebellion against the Protestant model. Rather, it was a consequence of the specialization and professionalization of higher education. When U.S. colleges and universities were small, professors were often generalists. The man who taught astronomy might also teach the New Testament, and the one who taught Hebrew perhaps taught geology. But when faculties grew and were divided into various science departments, then astronomers and geologists had all their time occupied by those subjects, and those who rendered religious education lost their mandate to be naturalists as well. Astronomers, for example, looked to other astronomers for inspiration and affirmation, rather than to ministers. The finest scientists of that time, the role models for U.S. scientists in the mid–nineteenth century, were Europeans, most of whom embraced the secular principles of the Enlightenment, which recognized an important role for theories and hypotheses. Although the secularization of U.S. science was punctuated by notable events like the founding of Cornell University in 1865 and Johns Hopkins in 1876, it was mostly a gradual digression from Protestantism, both subtle and seemingly benign (Marsden 1989:41).

This was so even in the case of Darwinism (Marsden 1989:34). The U.S. Protestants who in the early 1860s might have initially opposed Darwin's great work, *On the Origin of Species by Means of Natural Selection*, were occupied instead with issues of slavery, secession, and civil war, while those who endorsed it had the advantage of being led by Asa Gray, the Harvard botanist who knew Darwin well enough to be his personal emissary to U.S. scientists. Charles Hodge argued eloquently that Darwinism was a mortal threat to the doctrine of design, but the Evangelical Alliance, meeting in New York in 1873, accepted the theory that Protestants could interpret Darwinism—and accept it—in terms that were compatible with their theology (Marsden 1980:20). That was hardly a ringing endorsement of Darwinism, but it neverthless deflected Hodge's

18

call for a head-to-head conflict between evolution and creationism. As late as 1915, the seminal essays that established fundamentalist theology were distinctly more accommodationist than confrontational on the issue of Darwinism (McIver 1989:216–230; Marsden 1989:37).

Another factor that masked science's gradual separation from Protestantism was its selective base of social support. In the nineteenth century, scientists were the main disseminators of scientific knowledge (Burnham 1987:26). Direct knowledge of secular scientific values reached relatively small, sophisticated audiences that attended learned lectures and read intellectual periodicals (Rosenberg 1966:154). The widespread popularization of science in U.S. culture, when large audiences received large amounts of information about modern science (with or without a framework of secular Enlightenment values), began in the early twentieth century. One vital feature of popularization was the explicit recognition that, because science seemed to be the source of material prosperity, it must be an autonomous moral authority, independent of Protestantism. Science was generally reported to be the pronouncements of authorities, and it was widely accepted as "an absolute able to justify and motivate individual action" (Rosenberg 1966:136; and see Burnham 1987:239; Hollinger 1989). "As popular culture increasingly linked social progress to science, scientists found their intelligence and knowledge to be unchallenged and their opinions in great demand," and "scientific and engineering knowledge appeared to solve every problem, to supply an answer to every question" (LaFollette 1990:100,9).

And so three cultural models of science were circulating in this country by the early decades of the twentieth century. The first was the nineteenth-century Protestant model, which described scientific reasoning in terms of the Common Sense philosophy, Baconian empiricism, and the Princeton Theology. This reasoning, however, had become the lonely province of conservative Protestant theologians, but of few others. The second was the secular model of the European Enlightenment, explicitly grounded in rationalism and naturalism. Most U.S. scientists, especially those employed in higher education, had adopted this way of thinking, with the consequence that U.S.

19

science and science education were thoroughly evolutionist, whether Darwinian or Lamarckian, by 1900 (Larson 1985:chap. 1; Numbers 1992:chap. 1).

The third was the model that valued science as a source of consumer products. This model embraced no particular epistemology; it simply judged science in terms of tangible results. There was no end to the images, metaphors, and stereotypes depicting the belief that science deserves respect because it produces prosperity, but that short-circuit reasoning was distinctly more shallow and more trivial than, say, the Protestant model or the secular model. Usually this understanding is called American pragmatism, but because of its intellectual consequences for the understanding of scientific knowledge, I refer to it as the trivial model of science. Unlike the other two models, this one had little or no apparent logic to anchor the symbols of science—inventors and inventions; lab coats and stethoscopes; test tubes, centrifuges, and retorts—in any particular intellectual content. The connection between the substantive meaning of science and the popular symbols of science was so weak that the symbols could easily be borrowed, co-opted, or stolen for the benefit of ideologies, policies, and commodities that did not necessarily have anything to do with the substance of science. In short, the trivial model invited Americans to respect the symbols of science rather than to understand its content.

Coinciding with the ascendancy of the secular and trivial models of science was the emergence of fundamentalism. Conservative U.S. Protestantism took a turn toward pessimism in the late nineteenth century, when it seemed that the basic Christian message of salvation had been displaced by a concern that should have been secondary, namely, the "Social Gospel," which was much consumed by issues of poverty, justice, and hygiene. In reaction, conservative Christians attempted to reduce Protestant thought to the issue of salvation, and to anchor that issue in one source of truth, namely, revelation. During the first two decades of the twentieth century, the leaders of the fundamentalist movement developed a theological nexus of salvation and revelation that, not surprisingly, depended heavily upon the Princeton Theology. When fundamentalist thought was extended from theological questions to cultural matters in general, then all secular influences, including secular science, were diagnosed as threats to salvation and revelation.

Fundamentalism began as a *moral theory* about religion and culture, but in the 1920s it coalesced into a *moral crisis*. The pervasive disrespect for law that accompanied Prohibition; the irresponsible liberties of the Roaring Twenties and the Jazz Age; the profound disappointment that distinguished U.S. idealism in World War I (the war to end all wars; the war to make the world safe for democracy) from the greedy cynicism of the European victors: all these convinced fundamentalists that Armageddon was at hand. Although fundamentalism was actually a critique of secular culture in general, William Jennings Bryan and other conservative Protestants soon narrowed that critique down to the accusation that Darwinism was responsible for everything from German militarism to the supposed atheistic takeover of U.S. schools (Bryan 1922, 1925a, b, [1921] 1967; Ginger 1958; Numbers 1982).

Yet their views on science were hardly reconciled with the cultural realities of Americans' feelings about science. Were Bryan and his colleagues hostile to science? Yes and no. They were very much in favor of the nineteenth-century Protestant understanding of science. In fact they employed the principles of Common Sense and Baconian empiricism to criticize evolutionists for being unscientific (Larson 1985:45; Numbers 1982:539; Webb 1986). By the twentieth century, however, the U.S. scientific community, especially scientists in higher education, had embraced the epistemology that was no longer wedded to Protestant thought. In effect, then, the fundamentalist anti-evolutionism of the 1920s was friendly to nineteenth-century scientific values and hostile to twentieth-century scientific values. Consider these comments from anti-evolutionists of the 1920s:

- William Jennings Bryan: "Man is infinitely more than science; science, as well as the Sabbath, was made for man" (Larson 1985:46).
- Austin Peay, governor of Tennessee at the time of the Scopes trial, explained that the Butler Act, which outlawed

Scientific Authority

the teaching of Darwinism, was a "protest against an irreligious tendency to exalt so-called science, and deny the Bible in some schools" (Larson 1985:57).

- Thomas Stewart, prosecutor in the Scopes trial: "I say, bar the door, and not allow science to enter" when evolution threatens religion (Larson 1985:89).
- An unnamed state legislator in Mississippi: "Great scientists look at all things as material and they frequently lose sight of the spiritual. I'd rather have the leadership of one Christian mother than of all the scientists in the world" (Larson 1985:78).
- Billy Sunday on evolution: "When the word of God says one thing and scholarship says another, scholarship can go to hell!" (McLoughlin 1955:132).

To these we can add Bryan's well-known aphorism that he was more interested in the Rock of Ages than the age of rocks, and also that he feared a "scientific soviet" would "establish an oligarchy over the forty million American Christians" (Numbers 1982:538).

A hard difference between creationists and evolutionists had crystallized during the 1920s. The former were, literally, anti-evolutionists: regardless of whether they believed that the Holy Bible was free from error, they were certain that the idea of evolution was a source of immorality. As to whether evolution was scientific, the creationists could appeal to the Protestant standards of the previous century and conclude that evolution was unscientific, but by the secular standards of the twentieth century, the evolutionists could come to the opposite view. It is understandable that these two sets of values coexisted in the 1920s if one considers that the change from the former standards to the latter was gradual, subtle, recent, and, conseqently, not widely realized by fundamentalists or the rest of the U.S. public, for the reasons mentioned earlier. And, because fundamentalist anti-evolutionists depended on the scientific expertise of a pair of self-proclaimed scientists who were really irresponsible amateurs—that is, George McCready Price and Harry Rimmer (Numbers 1982, 1992)—they were especially ill-informed about the secular values in scientific thinking.

22

This is not to say that the secular model had successfully wrested any of the moral authority of science from the Protestant model. One of the principal goals in the defense strategy of Clarence Darrow and his associates at the Scopes trial was to educate the U.S. public regarding the rationalist and naturalistic values of the secular model (Ginger 1958:154–162), which presumes that the values well known to scientists were hardly known at all to the public.

Less than a century earlier, the leading American model of science was simultaneously Protestant, morally authoritative, and representative of the values of most naturalists. Now, however, those three cultural features had broken apart and had become three separate models. Thus the creation-evolution controversy of the 1920s was, among other things, an unsuccessful attempt by fundamentalists to recover the moral authority of science that they had recently lost, and also an unsuccessful attempt by evolutionists to claim that moral authority for their own secular scientific values. In fact it was captured by neither of those two, but rather by the third, the trivial model. Ideas and products were ennobled simply by dressing them in the symbols of science, irrespective of intellectual content or fidelity to scientific methods, whether Protestant or secular. Scientific respectability was up for grabs.

The three-way fragmentation of science's role in U.S. life continued in the decades after the Scopes trial. By the 1950s, conservative Protestant thinking about science was acutely pessimistic. Fundamentalists "had become almost thoroughly isolated and alienated from the dominant American scientific culture" (Marsden 1989), as evidenced by the ersatz science of Harry Rimmer and George McCready Price. Finally, in the late 1940s, some leaders of the American Scientific Affiliation insisted that secular scientific thinking had merit and credibility, regardless of its implications for Protestant theology (Numbers 1982:541). The Seventh-day Adventist church experimented with an open-minded attitude about secular science, but the experiment failed when the theological implications became alarming, at which point the church purged secular scientific values from its institutions (Numbers 1979). With the Adventists operating as protectors of creationist orthodoxy, one of the few expressions of anti-evolutionary sentiment to reach the general public came from Frank L. Marsh, an Adventist scientist, who complained in 1960 that evolution ought not to be taught as a proven fact (*New York Times*, 25 June 1960).

Conservative Protestant thinking about culture in general was likewise pessimistic. One theologian complained, in 1956, that words like "ethics" and "morals" were "almost totally absent from the working vocabulary of this generation [of young people]. Their lips would hardly know how to frame and utter these words." Most students at the Stevens Institute of Technology in Hoboken, New Jersey, admitted in 1960 that they had abandoned religious sentiments that conflicted with scientific values. The Billy Graham Crusade completed a hugely successful campaign in New York City in 1957, which in evangelical folklore became an exciting story of how God's favorite preacher survived a visit to Sodom and Gomorrah. When the *New York Times* surveyed religion in the New York area the next year, however, it erased the happy accomplishments of the evangelicals by demonstrating that the Billy Graham Crusade had left little lasting effect (*Newsweek*, 23 January 1956:59; *New York Times*, 26 December 1960, 21 June, 3, 21 July 1957, 26 January 1958).

By contrast, the institution of science enjoyed enormous optimism and prestige in those years. President Eisenhower honored the men and women of the scientifc community with a National Committee for the Development of Scientists and Engineers; while the president blessed the National Science Foundation with a 256 percent budget increase in 1956, a Methodist bishop consecrated scientific achievement by commenting that religion would supply it with moral purpose. When the nation learned in October 1957 that the Soviet sputnik was casually violating the heavens—*our* heavens, our *American* heavens—scientists, engineers, and technicians became the combat officers of the cold war. And less than a year after the first news of the sputnik, "the Pearl Harbor of technological war," our leaders congratulated themselves on scientific improvements; a mood that had begun as eloquent panic had drifted into caution against complacency. According to the most common platitude of that season, we had made great strides, but we still faced

24

great challenges (*New York Times*, 17 January, 7 September 1956, 20 October 1957).

The morality of the U.S. space program, however, was equivocal. Whereas the Soviets had supposedly threatened us treacherously when they launched their sputnik, they soon became our partners in joint ventures to conquer space and produce space technology. Within a period of about a year and a half, they had been our deadly enemies and our estimable rivals and our trustworthy partners. Similarly, atomic energy emerged as a popular symbol of modern science, but it, too, was enmeshed in ethical confusion. It gave us good bombs to fight bad enemies, who in turn had bad bombs. Their use by either us or our enemies would be terrible. Though atomic bombs were bad, atomic energy was good—a godsend, supposedly, that would provide cheap electricity for U.S. homes.

Good news about science included good news about evolution. In 1956, the Darwin Anniversary Committee led the scientific community through a belated centennial celebration of the voyage of the H.M.S. *Beagle*, and two years later Sir Julian Huxley, the high priest of evolutionary metaphysics, proposed a strange science of "human possibilities" that would guide human evolution. This was followed by a pair of exciting announcements in September and October 1959, when Louis Leakey announced that a fossil skull from the Olduvai Gorge in Tanganyika was at least 600,000 years old, and that it represented the first "true man." In November of that year evolutionary thought enjoyed an even grander climax, when the scientific community celebrated the centennial of the publication of Charles Darwin's *Origin of Species*, with Julian Huxley again proclaiming his ideas about a new theology based on evolutionary values (*New York Times*, 17 July 1958, 4 September, 8 October, 27 November 1959).

Those events seemed to prove that, in U.S. culture, evolutionary thought had triumphed over conservative Christian belief in creationism, which is to say that the secular model had displaced the Protestant model. Yet conditions were not that simple. In fact a peculiar paradox complicated this situation: the concept of evolution was practically absent from science education in the public schools. Many state and local education

25

officials desired desperately to protect their communities from the ridicule endured by Dayton, Tennessee, during the Scopes trial in 1925. Regardless of their personal beliefs about creation and evolution, they often felt that the best way to insulate their schools from this controversy was to have *neither* evolution nor creation in their classrooms. The publishers of science textbooks altered their products to conform to this attitude, so that evolutionary topics received only superficial treatment in textbooks. Explicit mention of evolution disappeared from science textbooks as a parade of euphemisms like "development" and "changing forms" shielded public education from plain knowledge of evolutionary science (Grabiner and Miller 1974; Skoog 1983). In one particularly outrageous editorial dissemblance, the publishers of *Modern Biology* twisted a statement by T. H. Huxley (grandfather of Julian) to imply the opposite of his original meaning. Huxley, the most important nineteenth-century popularizer of evolutionary ideas, had originally written that he lived quite contentedly without orthodox Christian religion because modern science gave him faith and comfort comparable to the benefits of formal religion. *Modern Biology* cited his comments on the benefits of formal religion but deleted the context of those comments, thereby leading biology students to believe that this most agnostic of agnostics had embraced orthodox Christianity (Gould 1982).

Several influences stoked publishers' fears of evolutionary topics. The southern and western states, where anti-evolutionary sentiment was most intense, had a disproportionately strong influence on the purchase of biology textbooks because their agricultural economies required widespread education in plant and animal life. Furthermore, these states made centralized decisions about which textbooks to subsidize, thereby consolidating their influence vis-à-vis the northern and eastern states, where officials dissipated their own influence by giving much authority to local school districts. Finally, the authors of these textbooks were high school teachers responding to their publishers' marketing requirements, not scientists presenting their colleagues' consensus (Grabiner and Miller 1974). Thus the scientific community was isolated from science education, at least in the matter of teaching evolution: "The evolutionists of the 1920s believed they had won a great victory in the Scopes trial. But as far as teaching biology in the high schools was concerned, they had not won; they had lost. Not only did they lose, but they did not even know they had lost" (Grabiner and Miller 1974:836).

Even if the idea of evolution had practically unlimited credibility among U.S. scientists, it was poorly understood and not necessarily believed by most of the public. And so it was not the secular model of science that won the hearts of most people in the middle decades of the twentieth century, but rather the trivial model (Burnham 1987:170–225; LaFollette 1990:18–44). Unlike the Protestant model and the secular model, the trivial model had no apparent intellectual content or structure. "Science in the realm of the popularizers changed from a coherent view of nature . . . into choppy, unconnected 'facts'," plus trivial accounts of consumer products and sensational reports of new technology (Burnham 1987:5, 170, 173, 226). "The idea that science stood for something" disintegrated (Burnham 1987:248). Because the popular symbols of science were not anchored in any coherent intellectual structure, they were vulnerable to being borrowed or co-opted so as to confer the moral authority of science on ideas and products that were not necessarily scientific.

This, then, was the general condition of the Protestant, secular, and trivial models of science just before creationism was resurrected in 1961: the trivial model of science was a hollow one, for it conjured scientific sanctification by displaying the symbols of science, engineering, and technology, but it was unrelated to the substantive content of those forms of knowledge. The main motifs of that model were space technology, atomic energy, and the consumer technology that yielded the U.S. standard of living. Meanwhile, conservative Christians were unable to influence scientific thought in favor of traditional values based on biblical morality, and they were greatly troubled by many kinds of social change. The scientific world was not actively or explicitly *anti*-Christian but rather considered traditional Christian morality irrelevant to science and so ignored it. Finally, the idea of evolution, like other features of secular

27

science, seemed generally modern and respectable, but few people understood its logic or its principles.

When the nation turned to science and technology to save it from the sputnik threat, the scientific community pointed out that getting serious about science included getting serious about evolution. The National Science Foundation commissioned the Biological Sciences Curriculum Study (BSCS) in Boulder, Colorado, to produce a series of biology textbooks that would include evolution prominently and explicitly. Beginning in the early 1960s, biology textbooks, led by the BSCS books, drastically reversed the thirty-five-year neglect of evolutionary topics (Moore 1976:192–193; Nelkin 1977:27–28). This was the last major affront to the opponents of evolutionary thought before the rise of modern creationism, for it meant that the logic and principles of secular science might actually be passed from scientists to large portions of the population.

Part Two
The National Creationist Movement

Three
The Renaissance of Creationism

A remarkable book that appeared in March 1961 redeemed anti-evolutionary thought from thirty-five years of obscurity and self-doubt. *The Genesis Flood*, by John C. Whitcomb, Jr., and Henry M. Morris (1961), reversed the painful popular assumptions that had plagued U.S. anti-evolutionism since the Scopes trial of 1925; namely, that science and conservative Protestantism stood against each other in philosophical conflict, that science regularly overcame religion because of its superior command of reality, and that naturalistic evolution ordinarily overcame biblical creationism just by acting out this simple script. By matching conservative Christian belief in biblical inerrancy with a medley of scientific idioms, this volume clothed the six-day Creation, Adam's Sin, the Fall of Man, and Noah's Flood in a garment of scientific respectability.*

Whitcomb and Morris challenged that chain of beliefs boldly and brilliantly. By adducing technical authority in support of biblical narrative, they wove together conservative Protestant

---

*Inerrancy is the belief that the Holy Bible is without error in whatever it addresses, but that humans sometimes make errors of interpretation. This is a slightly more flexible stance than infallibility, which became indefensible when it did not distinguish between scriptural error and human error. Inerrancy is also more flexible than biblical literalism. There is no room for metaphor in the latter, not even in the parables. Inerrancy favors literal interpretation in most cases, including the stories in Genesis, yet it admits that sometimes scripture speaks through figures of speech.

31

values and the American faith in science, creating one fabric of belief out of two. Hereafter, conservative Christians would not have to make agonizing choices between the scientific climate of the modern world and the pious spirit of biblical values. Now they could turn to Whitcomb, a theologian, and Morris, an engineer, to hear that science validates biblical belief, in general and in particular. Inspired by this novel departure from the science-versus-religion assumptions of the mid–twentieth century, Bible-believing Christians reorganized their faith in literal creation and their hostility to evolutionary thought. The product of this reorganization was the movement known today as scientific creationism.

As an attempt to reestablish the primacy of the nineteenth-century Protestant model by borrowing authority from the trivial model, scientific creationism was predicated on two cultural realities: (1) Although the Protestant model had lost its former primacy, the secular model did not monopolize the moral authority of science; the U.S. public, wherein moral authority ultimately resides, invested that authority in the trivial model instead of the secular model; and (2) the symbols of that authority could be easily looted from the trivial model, which has no particular intellectual structure in which the symbols of scientific authority are anchored. Thus it is not surprising that this version of creationism attributed much merit and credibility to science, provided that merit and credibility could be interpreted to reflect well on creationist thought.

In this chapter I recount the principal events since 1961 in three overlapping phases: first, the reemergence of militant creationism and the establishment of its organizational infrastructure between 1961 and 1972; second, a period of successful proselytizing and expansion from 1969 to about 1982; and third, a series of anticreationist counterattacks and legal obstructions since 1977.

The Rebirth of Creationism, 1961–1972

Henry M. Morris was a young engineer who rejected evolution and embraced creationism while teaching at Rice University in Houston, Texas, during the early 1940s (Morris 1984:93–96).

The Renaissance of Creationism

At that time he was a member of the American Scientific Affiliation (ASA), a broad-based evangelical group of scientists that included both strict creationists, who wanted their scientific findings to conform to biblical inerrancy, and theistic evolutionists, who found comfort in the less contentious idea that evolution occurred in nature within the context of God's spiritual plan. Henry Morris agreed with the strict creationists that the theistic evolutionists were compromisers. In 1943 he became attached to the flood geology theories of George McCready Price (ibid.:80). Price, an old-time agitator for strict creationism, had proposed in 1923 that Noah's Flood accounted for all geological formations and fossil deposits (Numbers 1982:540), thereby dismissing the chronological data at the heart of evolutionary research. When Morris went to the University of Minnesota in the late 1940s to earn a doctorate in engineering, he decided to specialize in hydraulics and geology "primarily because of their importance in the study of [Noah's Flood]" (Morris 1984:136).

Morris met John C. Whitcomb, Jr., a young theology student who shared his hardline views, at an ASA meeting in 1953 (ibid. 1984:146). The two worked together on an elaborate update of Price's theories, with Whitcomb supplying the theology, and Morris the science. The result of their collaboration was *The Genesis Flood*, published in March 1961. This book presented two complementary themes: first, that scientific evidence illuminated biblical belief; and second, that naturalistic phenomena could be explained by principles of biblical inerrancy. Thus, for example, *The Genesis Flood* stated that "the fossil record, no less than the present taxonomic classification system, and the nature of genetic mutation mechanism, shows exactly what the Bible teaches" (Whitcomb and Morris 1961:450). With this approach, creationism turned its back on the old supposition that scientific evidence negated biblical belief. In its place it offered a "scriptural framework for historical geology" that included geological details to decorate scriptural chronology.

Whitcomb and Morris made flood geology the central paradigm of the new creationism. Whereas the old-time creationism of the 1920s had been almost exclusively concerned with *human*

33

The National Creationist Movement

origins and evolutionary implications about human nature, the latter-day creationism of Whitcomb and Morris was centered on the idea of Noah's Flood and its implications for geochronology. The earlier creationists had feared the belief that humans were descended from apes, insisting instead that we had been made in God's literal image. While Whitcomb and Morris and their followers mentioned this issue occasionally, they concentrated on denouncing evolutionary chronologies that took for granted millions of years of earth history. Their answer on this matter was that the earth was very young, and that the Flood explained most earth history. This particular emphasis left no license to disagree about biblical chronology (Numbers 1982:538): God created the world in six literal twenty-four-hour days, not six figurative time periods; no events, whether geological or spiritual, occurred before those six days; Noah's Flood was a universal deluge, not a local event in some little corner of the Holy Land.

Conservative Christians received the book happily, greeting it with kind reviews in their magazines and extending many invitations to Morris to lecture, to preach, and to write more about the case for creation (Morris 1984:157–159). At that time, the early 1960s, Morris belonged to a small network of scientists who traded ideas like his. This group, known in creationist tradition as the Team of Ten, felt inspired by the book's reception to organize a formal scientific society to encourage creationist research. Since most scientific journals had explicitly anticreationist editorial standards, the first priority of the Team of Ten was to publish a journal that would be a platform for creationist research (Rusch 1982:1). They formed the Creation Research Society (CRS) in June 1963 and began publishing the *Creation Research Society Quarterly* the next year.

Another new creationist organization began in September 1963, when Rev. Walter Lang, a Lutheran minister in Caldwell, Idaho, founded the *Bible-Science Newsletter* "to stimulate an exchange of ideas on Bible-Science relationships" (Lang 1983). Lang was apparently acting independently of Morris and his circle at that time. In 1964, Lang expanded his organizational services by establishing the Bible-Science Association (BSA) as a lay counterpart to the Creation Research Society.

34

The Renaissance of Creationism

The founding of two more creationist groups in the early 1970s completed the basic infrastructure of the modern creationist movement. The Creation-Science Research Center (CSRC) was formed in 1970 in San Diego, California, to provide research, educational, and publishing services for the creationist movement. The Institute for Creation Research (ICR) was established in 1972, also in San Diego, when the founders of CSRC split that organization into two separate groups after disagreeing about motives and methods.

Creationism on the Offensive, 1969–1982

It seemed in 1969 that the creation-evolution disputes of the 1920s had finally been resolved by purging the last of the anti-evolution laws. The Tennessee legislature in 1967 had repealed the Butler Act, under which John Thomas Scopes had been prosecuted for teaching that humans were descended from apes. Arkansas's equivalent statute, the Rotenberry Act, was struck down by the U.S. Supreme Court in November 1968. An article in *Scientific American* in February 1969 happily reported those two events as The End of the Monkey War (DeCamp 1969). Ironically, however, a new, intense anti-evolutionism would burst forth in California that year. The issue was science education. If creationists could convince school boards that scientific creationism deserved a prominent place in science education, this would be a signal to the U.S. public that creationism had a strong claim to scientific authority, and it would also begin a process of producing scientific authority for creationism, as succeeding generations of schoolchildren learned to associate creationism with science.

California, like most other states, establishes policy for science education in its public schools by having the state Board of Education formulate a "science framework." This is neither a day-to-day lesson plan nor a week-by-week syllabus, for the Board members have no expertise to specify those details. Rather, the science framework is, supposedly, a general statement of principles and objectives from which science teachers and local school boards plan their science courses. Interest groups, however, sometimes interpret this document as an overly powerful

35

philosophical statement that determines all the particulars of science knowledge in the classrooms.

A state board of education like California's will revise its science framework occasionally to account for new developments in scientific knowledge, but not so frequently as to disrupt planning and textbook purchases. Typically, a state's science framework is revised about every five years, and California's was due for revision in 1969. A few years earlier, the Board of Education had routinely appointed an advisory panel of scientists to deliver a consensus on essential scientific principles. The advisory panel, in one of its dozens of statements, cited evolution as a major organizing principle of the life sciences. Vernon Grose, an aerospace engineer from Santa Barbara who belonged to a Pentecostal church, surprised the board, the panel, the science teachers, and even the state's creationist leaders by requesting a revision. Evolution, according to Grose's amendment, was to be described as less than certain, and creationism to be considered no less valid than evolution. For example, Grose suggested that "While the Bible and other philosophical treatises also mention creation, science has independently postulated the various theories of creation. Therefore, creation in scientific terms is not a religious or philosophic belief" (Moore 1976:196).

The advisory panel rejected most of Grose's suggestions, but the Board of Education thought favorably of them. Grose was dealing with a predominately conservative board: the superintendent of public instruction, an elected official who controlled the state's education bureaucracy, was Max Rafferty, one of the most conservative education officials in the nation. Many of the board members had been appointed by the governor, Ronald Reagan. The chair of the Board of Education, for example, was John L. Ford, a physician who belonged to the Seventh-day Adventist Church, one of the most fundamentalist denominations in the country.

During a protracted three-year disagreement over Grose's proposals, nineteen Nobel laureates in California asked the board not to include creationism in the science framework. The scientific advisory panel argued that creationism, as theology or philosophy, was inappropriate for science courses. Creationists

36

The Renaissance of Creationism

countered by attacking the idea of evolution, suggesting that "random selection" in evolution implied "that the origin of the world itself is a matter of accident or chance." This particular allegation was noteworthy, for it revealed a major change in the moral basis of anti-evolutionism. The anti-evolutionism of the 1920s had complained that the idea of evolution through natural selection was too deterministic, that it denied humans their free will as moral agents, but the version that emerged in the California dispute emphasized *too much* freedom in evolutionary philosophy, more or less equating evolution with total randomness: if people believe randomness is good and natural, they will feel free to indulge without restraint in selfish, immoral behavior. Both objections, one contradicting the other, can be found in the earlier and the later versions of creationism, but the lack of freedom was a major theme earlier, while excess of freedom became a major complaint in modern creationism (*New York Times*, 10 November, 12 December 1972).

The Board of Education endorsed Vernon Grose's revisions, but it squabbled with its own textbook commission for several years before the new science framework altered textbook purchases. Late in 1972, board chairman John Ford resolved the controversy with a compromise: the board agreed that evolution was a theory, not a fact; dogmatic statements about evolution and other matters were to be removed from the science curriculum and replaced by conditional statements; speculations about ultimate causes were to be considered nonscientific; and, finally, creationism, as a *philosophy* of origins, could be taught in *social science* courses (*New York Times*, 15 November, 15 December 1972, 13 January, 10 March 1973).

After Ronald Reagan left office in 1975, the next governor, Edmund G. ("Jerry") Brown, Jr., diluted the conservative influence on the state Board of Education by adding numerous appointees. The newly composed board eventually dropped the idea of teaching creationism in social science but retained the 1972 caveats on evolution and science (*Christianity Today*, 7 November 1980:67; *Science*, 20 March 1981:1332).

Vernon Grose's revisions in the California science framework signaled a major change in creationist political strategy. The anti-evolution laws of the 1920s had bluntly outlawed the

37

The National Creationist Movement

teaching of evolution, at that time defined as the theory of human descent from apes or monkeys, on the grounds that it contradicted the Holy Bible. Tennessee's Butler Act of 1925 was typical of those laws: "It shall be unlawful . . . to teach any theory that denies the Divine Creation of man as taught in the Bible, and to teach instead that man is descended from a lower order of animals."

In the modern strategy, however, creationists requested that if the public schools taught evolution, they should teach creationism, in equivalent proportions. The first such formal legislation, passed in Tennessee only six years after the repeal of the Butler Act, required "equal attention and emphasis to other ideas of creation, including those contained in the Bible." Evolution was to be balanced with *biblical* creationism, with no mention of *scientific* creationism. A Tennessee court noted the law's sectarian substance and ruled it unconstitutional on First Amendment grounds (*New York Times*, 9 May 1973, 11 September 1974).

In addition, creationists achieved notable successes in changing policies that did not require formal legislation, such as decisions by state and local boards of education. An anthology of articles by the Institute for Creation Research referred to the years 1971–1980 as "the decade of creation" because of the "explosive growth [that] has occurred primarily in the just-completed decade of the seventies" (Morris and Rohrer 1981:5). The creationist movement enjoyed victories in textbook selections or curriculum guidelines in Texas, Georgia, and Alabama, plus local successes in Columbus, Ohio; Dallas, Texas; Kanahwa County, West Virginia; Pulaski County, Arkansas; Tampa, Florida; and Melville School District, Missouri (*Christian Century*, 2 March 1977:188–189; Gish and Rohrer 1978:80–82; Nelkin 1973:95; Newell 1974; *New York Times*, 28 January 1977, 10 March 1981; *News & Observer* [Raleigh, N.C.], 27 February 1984; *Science*, 30 March 1984:1373; Skow 1981; *Southern Anthropologist*, May 1981:2).

Through 1983 and into 1984, an educational play about evolution called *Dandelion* had numerous bookings canceled in Connecticut, Florida, Georgia, and New York. The Paperback Players, an educational troupe based in New York City, had

38

been performing this piece in elementary schools for about sixteen years. The play contained simple humorous scenes about evolution, including a tadpole jumping out of the water to become a frog, and a primate standing up to become bipedal. Andy Lyman, a reporter with National Public Radio in New York, investigated the objections to *Dandelion*. Some educators were practicing self-censorship, he thought, without direct pressure from parents, preachers, or others. Lyman told me in a telephone interview in December 1984 that the authorities were becoming "upset without seeing the play. . . . No one has come out and been forthright [in their objections]. Evolution is a touchy subject. It's the suggestion of the climate which has people scared."

The creationist cause received a pleasant surprise in September 1980, when presidential candidate Ronald Reagan told a meeting of conservative Christian leaders that evolution

> is a theory, it is a scientific theory only, and it has in recent years been challenged in the world of science and is not yet believed in the scientific community to be as infallible as it once was believed. But if it was going to be taught in the schools, then I think that also the Biblical theory of creation, which is not a theory but the Biblical story of creation, should also be taught. . . . I have a great many questions about [evolution]. I think that recent discoveries down through the years have pointed up great flaws in it. (*Science*, 12 September 1980:1214).

Afterwards, Reagan never publicly mentioned the creation-evolution issue, and his administration never had any explicit policy on it. The only direct mention came from Robert ("Dr. Bob") Billings, a middle-level appointee in the U.S. Department of Education. Billings had previously been a theologian, an athletic coach, and a political coordinator for Moral Majority. While in the Department of Education, he commented that he'd like to see the "teaching of divine creation alongside the theory of evolution" in the public schools (*News & Observer*, 19 July 1981).

The year 1981 was a wonderful one for scientific creationism.

The National Creationist Movement

The case of *Segraves v. California* came to trial in March, as the Creation-Science Research Center sued to compel the California State Board of Education to protect thirteen-year-old Kasey Segraves, son of one of the CSRC directors, from being taught evolution in his science class. Judge Irving Perluss rejected all of the plaintiff's substantive points except to reaffirm the policy that evolution should not be taught dogmatically (*Newsweek*, 16 March 1981:67). The trial nonetheless evoked sensational interest in the creation-evolution controversy because the national press described the case as a reprise of the Scopes trial.

Also in March 1981, the Arkansas legislature passed the first of the modern laws mandating equal time for creation-science (as distinguished from biblical creationism) whenever evolution is taught. In the language of Act 590:

> "Creation-science" means the scientific evidences for creation and inferences from those scientific evidences. Creation-science includes the scientific evidences and related inferences that indicate: (1) Sudden creation of the universe, energy, and life from nothing; (2) The insufficiency of mutation and natural selection in bringing about development of all living kinds from a single organism; (3) Changes only within fixed limits of originally created kinds of plants and animals; (4) Separate ancestry for man and apes; (5) Explanation of the earth's geology by catastrophism, including the occurrence of a worldwide flood; and (6) A relatively recent inception of the earth and living kinds.

Also in 1981, Louisiana passed a balanced-treatment law, but it used an ambiguous generic definition: " 'creation-science' means the scientific evidences for creation and inferences from those scientific evidences."

A debate at Liberty Baptist College in Lynchburg, Virginia, was yet another happy 1981 event for creationists. On 13 October, Duane T. Gish of the Institute for Creation Research met Russell Doolittle of the University of California, San Diego, in a televised creation-evolution debate. All accounts agreed

40

that Gish was the winner. The scientific community tried to explain Gish's success in terms of timing and polish, not substance or logic, but the prospect of seeing this event on television demoralized many. "People were appalled by it," according to Porter Kier, a paleontologist at the Smithsonian. "The creationists are well practiced in this kind of presentation. Scientists are not" (*Science*, 6 November 1981:635; and see *Christianity Today*, 20 November 1981:43–45).

The last major creationist victory of 1981 came in November. A nationwide poll conducted for the Associated Press and ABC News asked 1,598 adults whether public schools should offer only "the theory of evolution," only "the biblical theory of creation," or both: 76 percent wanted both, and another 10 percent wanted creation only. A mere 8 percent thought that only evolution was appropriate (*News & Observer*, 19 November 1981).

Anticreationist Counterattacks, 1977–1987

Opponents of creationism won a handful of skirmishes in the mid-1970s. The National Science Foundation successfully defended itself against a lawsuit charging that unless NSF balanced its comments on evolution with equal credibility for creationism in its textbook project, the Biological Sciences Curriculum Study, the U.S. government would be "establishing secular humanism as the official religion of the United States" by its singular emphasis on evolution. The U.S. District Court dismissed that case in 1973, and the U.S. Supreme Court did the same in 1975. In a similar confrontation, Dale Crowley of the National Bible Knowledge Association sued the Smithsonian Institution in 1978 concerning an exhibit on evolution at the museum. Crowley said that to favor evolution while ignoring creationism amounted to an establishment of religion in violation of the First Amendment, but the U.S. District Court rejected his argument (Nelkin 1982:100–101).

By 1980 a pattern had become clear. Creationists could persuade many school boards to include creationism in their science curricula and textbooks, but their enemies could persuade the courts that those decisions violated the Establishment Clause of the U.S. Constitution ("Congress shall make no law respecting an establishment of religion").

The Arkansas case of 1981 presented the sharpest and clearest exercise in that kind of legal conflict. The Institute for Creation Research had decided in the early 1970s to concentrate on a grass-roots strategy of persuading local school boards, rather than state legislatures, of the merits of creationism, partly because legislative action would draw sensational attention and withering criticism. ICR formulated a draft resolution recommending "balanced treatment" for evolution and scientific creationism, to submit to local school boards. This plan was working well with the Pulaski County (Little Rock) Board of Education in Arkansas in early 1981. Larry Fisher, a math teacher, and W. A. Blount, a conservative preacher, shepherded their version of the draft resolution through the board's various conferences and hearings. Family Life America under God (FLAG), a local fundamentalist citizens' group, and the Greater Little Rock Evangelical Fellowship, a network of conservative preachers, provided popular support.

As they gained confidence from their success with the local board, Blount and the resolution's other supporters in the state decided—apparently without consulting ICR—to propose the policy to the Arkansas legislature. State Senator James Holsted steered the bill through the last-minute chaos of the legislature on the eve of its adjournment, protecting the bill from all but about fifteen minutes of hearings on its merits. Act 590, "An Act to Require Balanced Treatment of Creation-Science and Evolution-Science in Public Schools," became law in March 1981, with only a handful of votes against it.

The act had two notable features. First, the balanced treatment was to be very finely balanced. It applied to

> classroom lectures taken as a whole for each course, in textbook materials taken as a whole for each course, in library materials taken as a whole for the sciences and taken as a whole for the humanities, and in other educational programs in public schools, to the extent that such lectures, textbooks, library materials, or ed-

ucational programs deal in any way with the subject
of the origin of man, life, the earth, or the universe.

In other words, educators had almost no discretion in
weighing the scientific merits of creationism, which was to receive credibility precisely equivalent to that of evolution in a
broad range of courses and resources.

Second, Act 590 supplied a legal definition of creation-
science. To insure that creationism was taught as science, neither
a vague statement of discontent with evolution nor a reference
to Genesis would work. Lawyers, teachers, principals, and
schoolchildren would have to know just what creation-science
was. Thus the precise description of creation-science, cited earlier. These two features made possible a legal challenge by the
American Civil Liberties Union. The ACLU appealed to a three-
part test that the federal courts had developed for assessing
whether an act was unconstitutional on First Amendment grounds
prohibiting "establishment of religion." In the words of Chief
Justice Warren Burger, author of the *Lemon v. Kurtzman* opinion
of 1970, "First, the statute must have a secular legislative purpose; second, its principal or primary effect must be one that
neither advances nor inhibits religion . . . ; finally, the statute
must not foster 'an excessive government entanglement with
religion' " (403 US 612). The ACLU complaint challenged the
Arkansas law on all three criteria.

The judge concurred with all of the ACLU's important points.
After reviewing an enormous body of information that included
pretrial depositions, private correspondence among creationist
lobbyists, many items of creationist literature, and testimony
from scientists, educators, ministers, and philosophers, Judge
William R. Overton wrote an opinion affirming most forcefully
that Act 590 violated *each* of the three Establishment Clause
tests. His criticism of Act 590, in fact, was even sharper than
the ACLU's complaint. First, to establish the point that the act
had a sectarian purpose, Overton wrote that the correspondence
of Paul Ellwanger, one of the creationist lobbyists, "shows an
awareness that Act 590 is a religious crusade, coupled with a
desire to conceal this fact," and, that "Senator Holsted's sponsorship and lobbying efforts in behalf of the Act were motivated

solely by his religious beliefs and desire to see the Biblical version of creation taught in the public schools" (Overton 1982).

On this critical constitutional matter, the act lacked credible evidence of any secular motivation simply because it had been rushed through the legislature without serious hearings or substantive deliberations. Since it had no legislative history, Judge Overton could refer by default to the lobbyists behind the bill to assess motivations: "The author of the Act had publicly proclaimed the sectarian purpose of the proposal. The Arkansas residents who sought legislative sponsorship of the bill did so for a purely sectarian purpose. . . . It was simply and purely an effort to introduce the Biblical version of creation into the public school curricula. The only inference which can be drawn from these circumstances is that the Act was passed with the specific purpose by the General Assembly of advancing religion" (ibid.).

Next, to decide whether the act had the effect of advancing religion, Overton referred to its definition of creation-science: "The evidence establishes that the definition of 'creation science' contained in 4(a) has as its unmentioned reference the first 11 chapters of the Book of Genesis. . . . The concepts of 4(a) are the literal Fundamentalists' view of Genesis." Further, "the ideas of 4(a)1 are not merely similar to the literal interpretation of Genesis; they are identical and parallel to no other story of creation." In a footnote, Overton wrote that: "Senator Holsted testified that he holds to a literal interpretation of the Bible; that the bill was compatible with his religious beliefs; that the bill does favor the position of literalists"(ibid.).

After that, the judge dealt with the charge that the Arkansas law would lead to state involvement in religious matters: "State entanglement with religion is inevitable under Act 590. Involvement of the State in screening texts for impermissible religious references will require State officials to make delicate religious judgments. . . . These continuing involvements of State officials in questions and issues of religion create an excessive and prohibited entanglement with religion" (ibid.).

Thus, the bill's extremely precise definition and its balance of creation-science and evolution-science had become its fatal flaws.

44

The Renaissance of Creationism

The creationists' witnesses did not serve their cause well. Norman Geisler of Dallas Theological Seminary, a thoughtful man with a respectable Ph.D., earned headlines for his belief that UFOs are satanic deceivers. Margaret Helder of Edmonton, Alberta, had to admit she had not taught science classes since 1974. William Scott Morrow of Wofford College in Spartanburg, South Carolina, alleged that evolutionists suppress scientific articles supporting creationism, but he was unable to substantiate even one case of this practice. Chandra Wickramasinghe of the University of Wales complained about scientists' "indoctrination in Darwinism," but he also denounced creation-science as "claptrap" (*New York Times*, 15 December 1981; *Washington Post*, 17 December 1981).

The litigating of the Louisiana law was also painful to creationists. The legislature of that state had defined creation-science in the most ambiguous terms possible, but the federal judge in that case inferred that creation-science in the Louisiana law was the same thing as creation-science in the Arkansas law. On the basis of its connections with the Arkansas case, U.S. District Court Judge Adrian Duplantier ruled that, "because it promotes the beliefs of some theistic sects to the detriment of others, the [Louisiana] statute violates the fundamental First Amendment principle . . . that a state must be neutral in its treatment of religions" (*Science* 25 January 1985:395). Judge Duplantier resolved this case in a summary decision that assumed that the question of whether creationism was a science had been satisfactorily adjudicated in the Arkansas case. This left only one matter of law to consider, that is, whether the Louisiana case was sufficiently similar to the Arkansas case to reaffirm the Arkansas decision. It was, thought Duplantier. He ruled two times that it was unconstitutional, once in November 1982 and again in January 1985. The Fifth U.S. Circuit Court of Appeals upheld Duplantier in a decision by three judges in July 1985 and again in January 1986, the second time with an eight-to-seven vote. Still, the law kept rising from these setbacks as the state attorney general, William Guste, sought a new hearing at a higher level after each defeat.

The U.S. Supreme Court surprised creationists and evolutionists alike by agreeing in May 1986 to adjudicate the

45

Louisiana case. At the time it seemed peculiar, for this case threatened to disrupt the centrist formula for First Amendment church-state separation that the Court has hammered out very carefully in recent years, for example, by striking down a series of laws encouraging group prayer in public schools while allowing "moment-of-silence" laws. The Court's decisions in these and other religion cases constituted a policy that was firm but flexible, keeping church and state apart without humiliating either. Considering this legal context, it was hard to see why the Supreme Court would want to hear the Louisiana case. If this was another routine consideration of church-state issues, why not let the lower-court decisions stand? And if the Court intended a major departure from recent precedents, what had been the point of developing those precedents so carefully?

The Supreme Court heard the oral arguments on 10 December 1986. As expected, Jay Topkis of the American Civil Liberties Union proposed that the Louisiana law was essentially sectarian, having no scientific merit, while Wendell Bird, the creationist legal strategist, said that the law's scientific merits greatly outweighed whatever religious attributes there were (*Science*, 2 January 1987:22). Since Judge Duplantier had originally ruled on the case in summary decision, Bird's goal was for the Supreme Court to order a full trial for the Louisiana law. This the creationists would have enjoyed: the opportunity to tell Judge Duplantier all their theories, models, evidences, and hypotheses, and to deluge him with books, tracts, filmstrips, slideshows, and videotapes. They had assembled a thousand pages of depositions and affidavits for this case, and they resented Duplantier for ruling on issues of law without considering their factual evidence.

The U.S. Supreme Court announced its decision on 19 June 1987. Seven of the nine justices agreed that Louisiana's balanced-treatment law was unconstitutional in the sense that the legislature's intention had been sectarian, that is,

> to advance the religious viewpoint that a supernatural being created humankind. The term "creation-science" was defined as embracing this particular religious doctrine by those responsible for the passage of the Cre-

46

ationism Act. . . . The legislative history documents that the Act's primary purpose was to change the science curriculum of public schools in order to provide persuasive advantage to a particular religious doctrine that rejects the factual basis of evolution entirely. (*New York Times*, 20 June 1987)

Regarding the claims of academic freedom by which the Louisiana lawmakers had sought to justify the bill, Supreme Court Justice William J. Brennan, Jr., wrote that,

if the Louisiana Legislature's purpose was solely to maximize the comprehensiveness of science instruction, it would have encouraged the teaching of all scientific theories about the origins of humankind. But under the Act's requirements, teachers who were once free to teach any and all facets of this subject are now unable to do so. . . . Thus we agree with the Court of Appeals' conclusion that the Act does not serve to protect academic freedom, but has the distinctly different purpose of discrediting "evolution by counterbalancing its teaching at every turn with the teaching of creation science." (55 LW 4860)

This decision wounded creationism in many ways. In addition to the immediate effect of nullifying Louisiana's creationist initiative, it dissuaded other state legislatures from enacting similar laws. Furthermore, the Louisiana decision became a useful tool for anticreationists who wanted to challenge local school board policies that accommodated scientific creationism. The Supreme Court judgment did not categorically prohibit creationism, but it placed a burden of proof on school boards to demonstrate that they had no religious motivations behind their creationist policies.

As he watched the progress of the Arkansas and Louisiana cases, Henry Morris observed that "the number and virulence of anti-creationist articles [had] accelerated rapidly" since 1977 (Morris 1984:313). In a 1981 letter he warned that: "the humanistic leadership of the educational and scientific establishments

47

is organizing for an all-out attack on the creationist movement, especially the Institute for Creation Research, which they have recognized as the key influence. A number of important meetings have been held, articles written and strategies developed for countering the growing creationist threat to the evolutionary monopoly over American education" (ICR cover letter, *Acts & Facts*, July 1981).

Anticreationists enjoyed successes in 1982 in New York City, Tennessee, and Georgia, where they defeated or reversed policies to include creationism in public school science education (*News & Observer*, 21 November, 28 February 1982; *Science*, 22 January 1982:380; *Scientific American*, September 1982:105–106; *Washington Post*, 24 November 1984). In the fall of 1984, some leading anticreationists were able to say that "it looks as if the tide has turned." Citing the Texas State Board of Education decision to reverse its creationist policies, Wayne Moyer of People for the American Way said, "For the first time, we felt that science was on the inside and creationism was on the outside" (*Washington Post*, 24 November 1984).

Creationism, Past and Future

Despite the confidence of the anticreationists, the modern form of creationism is here to stay. Even though almost every major scientific organization in the country has denounced creationism (McCollister 1989), the movement has convinced many millions of Americans that the mantle of scientific authority rests comfortably on the shoulders of creation-science. Public opinion polls consistently report that more than 40 percent of adults surveyed agree that creationism is as scientifically credible as evolution, and that it deserves a place in the science education curriculum (Eve and Harrold 1991:3–4; 87–88, chap. 3). Furthermore, it has a robust infrastructure of institutions like ICR and CRS that generate new lessons about creation, disseminate those lessons to fundamentalist schools and churches, and persuade their followers to conform to one particular line of interpretation.

Much of the appeal of modern creationism is made possible by changes in moral reasoning between the current form and

48

The Renaissance of Creationism

its ruder antecedents of the 1920s. Three shifts of moral ground are especially important. First, the basic moral complaint about the meaning of evolution has changed from fear of determinism, allowing no free will, to fear of randomness, allegedly permitting too much freedom of thought and action. William Jennings Bryan and his associates argued that evolutionary thought led people to believe that they were prisoners of biological history—"if people think they're descended from animals, then they'll behave like animals"—and therefore had no freedom to choose between good and evil. It is common for people to fear that powerful forces determine their lives; when this kind of fear is translated into a biological metaphor, it is not surprising that evolution would be interpreted by its enemies in terms of determinism.

The newer form of creationism, however, inverts the moral critique of evolution by defining it in terms of randomness. The modern critique edits out deterministic features such as adaptation through natural selection and connects evolution with immorality by charging that if people believe human beings arise through purely random events, their behavior will be random and disorderly. Creationists feel that the rise in disorderly displays of sexual freedom demonstrates the validity of this argument. Neither the earlier nor the current critique does justice to the concept of evolution, which balances both random and deterministic features, but each misdefines evolution in a radically different way. Creationism's moral critique of evolution has taken a 180-degree turn since the 1920s.

The second major change in creationist thought concerns the choice of the primary issues that clarify the difference between evolution and creation. In the 1920s, Bryan and other creationists focused narrowly on the question of human origins. As far as they were concerned, evolution meant human evolution only. Bryan said that he did not care whether plants and animals had evolved, provided human origins were protected from the theory of evolution (Bryan 1922). The biblical answer to evolution was Gen. 1:27: "God created man in His own image." But in modern creationism, Henry Morris and his colleagues at the Institute for Creation Research have made flood geology their central paradigm, as they assert that Noah's Flood is responsible for the surface of the earth as we see it today and for the fossil record embedded therein. Consequently, they dwell much more on matters like stratigraphy and geochronology than on human origins. Ironically, this approach shifts the scriptural focus away from the two creation accounts in Genesis 1 and 2, emphasizing instead chapters 6 through 11, which tell the story of Noah, his ark, and the Flood. Both versions of creationism, old-time and contemporary, address human origins and Noah's Flood, but their respective emphases are quite distinct. The one reduced the origin of species to the origin of the *human* species, and the other confuses the history of life with the geological evidence from which the history of life is inferred.

The third major change in creationist thought is the most important: the question of relations between science and religion. Because the nineteenth-century Protestant model of science had been eclipsed by the secular and trivial models, the old-time creationism of the 1920s seemed to be trapped on the losing side in a simple conflict framed as science versus religion. The brilliant innovation of modern creationism is that it refuses to accept the premise that conservative evangelical Protestantism must be at odds with modern scientific thought. Rather, it assumes that science serves religion by providing technical corroboration for biblical belief. In an age obsessed with scientific sanctification, the creationist movement provides great comfort to many conservative Christians by attributing scientific credibility to biblical beliefs and, conversely, by using simple biblical idioms to decipher scientific dilemmas. Some people welcome creationism just for its power to make life difficult for their enemies, including liberals, atheists, and humanists, but many others appreciate it for its double-barreled epistemology in which science is said to support the Bible as the Bible guides science. With this formula, scientific creationism has earned itself an important place in the fundamentalist schools of our nation.

At the same time, however, modern creationism has limits to its influence, and it discovers these limits painfully. Whenever the creation-evolution controversy comes to litigation, the Establishment Clause of the Constitution cripples the creationist case. Sympathetic educators can accommodate creationism in local decisions or unofficial policies, but constitutional obstacles

The Renaissance of Creationism

make it impossible to institutionalize creationism as permanent policy. Each of creationism's initial victories in California, Texas, Arkansas, and Louisiana has since become a bitter defeat.

Another problem plaguing creationism is the matter of public support. The arcane technical details of evolution and scientific creationism bore many a nonscientist. The issue of origins requires enormous popular interest just to nurture modest active support for each side's advocates. While 1981 and 1982 were great years for public controversy, popular interest has since declined. It may be that the buildup to the Arkansas trial, the trial itself, the judge's decision, and the immediate reactions to the decision were enough for most folks. While the rapid fall in public interest hurt both evolution and creationism, creationism suffered more because it depended so much more on mobilizing public support.

A third problem is creationism's position within the broad constellation of New Religious Right concerns. Creationism never gained the intense interest that surrounds the issues of abortion and public school prayer. Creationism, like loyalty to Taiwan or fear of rock music, has been more a sideshow. Although it may rise from time to time in popularity, the web of legal principles, scientific opinions, and moral crusades that surrounds modern creationism is too dense and too difficult to permit it to become the single most important concern of the nation's conservative Christians.

Having concluded that, it should be recognized that scientific creationism has earned itself a permanent role in the nation's fundamentalist schools, where it lets parents believe that their children can learn about the natural world without having to learn about evolution. It has a comparable function in the conservative Protestant churches that those parents attend, where it asserts that the sanctifying authority of modern science corroborates the details of Bible stories.

51

Four
Moral Interpretations of Evolution

Creationism is a moral theory that the idea of evolution is intimately involved in immorality, as cause or effect or both. This view also implies that one can repudiate immorality by adhering to a literal belief in creation. That premise, however, is a very general contention, and it is broadcast to a diffuse audience of believers and potential believers, variously labeled fundamentalists, evangelicals, born-again Christians, Bible-believing Christians, or Moral Majoritarians. These names have imprecise and overlapping meanings, for the groups themselves have imprecise boundaries and overlapping memberships. As a consequence, the general formula that "evolution equals immorality" is subject to multiple moral interpretations, depending on how it touches different peoples' understandings of morality and immorality.

To make some sense of the many anti-evolutionary understandings of evolution, it is useful to concentrate on four kinds of sectarian groups, each with its own creationists and potential creationists, and then to see how and why each of these factions embraces a different moral interpretation of evolution:

  1. *The New Religious Right*, that is, conservative Christians who have become politically organized on a religious basis to enforce moral standards, especially sexual morality, in public life. The New Religious Right judges an issue like the creation-evolution controversy according to its ultimate implications for moral order.

2. *The inerrancy Baptists,* who constitute the conservative wing of the Southern Baptist Convention. While they share many moral sentiments with the New Religious Right, their single most urgent concern is to defend the axiom that the Holy Bible is without error. This consideration governs their position on creation and evolution, so that their understanding of evolution lacks the broad moral critique of U.S. culture that the New Religious Right attaches to the idea of evolution.

3. *The evangelical center,* a body of pious Christians who have conservative social and moral sentiments, but who nevertheless concentrate almost exclusively on the mission of converting others to Christ. They ordinarily avoid controversial topics that could taint either their broad appeal or their simple message of personal salvation. Unhappily for them, the creation-evolution dispute is such a topic. They appreciate the religious basis of the creationist position, but they dread being identified with its divisive partisanship.

4. *The apocalyptic separatists* are Jehovah's Witnesses and the Worldwide Church of God, a pair of sects so alienated from mainstream U.S. culture that they disdain science and reject the idea that the Bible needs scientific corroboration to confirm its own authenticity. This position sets the apocalyptic separatists apart from the scientific creationist movement.

The New Religous Right

Henry Morris and the Institute for Creation Research have long enjoyed close connections with two of the most important leaders of the New Religious Right: Rev. Tim LaHaye of El Cajon, California, and Rev. Jerry Falwell of Lynchburg, Virginia. In 1970, Morris left his position as a professor of engineering at Virginia Polytechnic Institute in order to establish a research center for creationism that would be a "great center of Biblical truth," and he wanted it to be nested within a conservative Christian college (Morris 1984:13). At the same time, LaHaye, pastor of an Independent Baptist church named Scott Memorial, was expanding his network of Christian academies to include a

53

college. LaHaye and Morris together founded Christian Heritage College (CHC), with Morris's research center as its intellectual hub.

LaHaye's great ideological contribution to the New Religious Right was to write *The Battle for the Mind* (1980), a polemic that identified "secular humanism as the arch-enemy of the New Christian Right" (Hadden and Swann 1981:85), thereby giving fundamentalist Christians a seemingly tangible enemy upon which to focus their hostility. When Falwell formed Moral Majority in 1979, LaHaye became its California chair. Later, LaHaye established the Council for National Policy, a network linking conservative Christian leaders with right-wing politicians and financiers (*Newsweek*, 6 July 1981:48). More recently, he expanded his ministries to include the American Coalition for Traditional Values (a right-wing mailing-list lobby) and Concerned Women for America, a similar organization led by his wife, Beverly LaHaye.

Morris and his creationist institute were linked intimately with LaHaye's college. From 1970 to 1978, Morris was the school's vice-president for academic affairs, in which role he composed a fourteen-point faculty doctrinal statement emphasizing the centrality of creationism, "with all curricula to be founded on creationism and full Biblical authority" (Morris 1984:222). The doctrinal document specified that the creation account of Gen. 1:1–2:3 would be "foundational to the understanding of every fact and phenomenon in the created universe," and that evolutionary thought "in any form" was to be rejected categorically (Morris 1984:355–359). "All the courses and curricula of the College [reflect] the creationist philosophy of [the Institute for Creation Research]" (Morris and Rohrer 1981:297); sophomores at CHC took six hours of scientific creationism from faculty with joint ICR-CHC appointments (Morris 1984:226; Morris and Rohrer 1981:304–306). ICR reported that "Dr. LaHaye is an enthusiastic supporter of ICR and gives outstanding over-all direction to the combined ministries of both the College and the Institute" (Morris and Gish 1976:171). When LaHaye left to devote his attention to another project, Henry Morris served as president of CHC from 1978 to 1980. In effect, a seamless fabric of beliefs and associations connected

54

the Institute for Creation Research with Christian Heritage College. Both LaHaye and ICR are now formally separated from the college, but LaHaye's generosity during ICR's early years nurtured creationism's most important organization in its formative period.

Meanwhile, in Virginia, Rev. Jerry Falwell promoted scientific creationism at his school (then known as Liberty Baptist College; it changed its name to Liberty University in April 1985). In September 1978, Henry Morris and Duane T. Gish, associate director of ICR, were the featured speakers at Falwell's Thomas Road Bible Conference in Lynchburg. In the ICR newsletter it was noted that "It is significant that a commitment to literal recent creationism is required of all faculty members at Liberty Baptist College, Thomas Road Bible Institute, and Thomas Road Baptist Seminary. A number of ICR publications are used as textbooks in courses at these schools" (Morris & Rohrer 1981:272). (The Bible institute and the Baptist seminary are units of Liberty University.)

The next year, Falwell told his faculty, "I want you to have all the academic freedom you want, as long as you wind up saying the Bible account [of creation] is true and all others are not." Early in 1981, Morris preached part of a service at Falwell's Thomas Road Baptist Church; later this was broadcast on Falwell's weekly television program, the "Old-Time Gospel Hour." Falwell also hosted the October 1981 debate between Gish and Russell Doolittle and likewise televised that on the "Old-Time Gospel Hour." Falwell denounced evolution—"there is not one shred of scientific evidence to support it"—in his *Penthouse* interview of March 1981. At a rally in Raleigh, North Carolina, in September of that year, he told his audience that "evolution is the cardinal doctrine of humanism," and "I want to see creation taught alongside evolution in the public schools." Henry Morris recognized the creationist stance of the "Old-Time Gospel Hour" and Liberty Baptist College as "committed to literal creationism." When Morris was president of the Trans-National Association of Christian Schools, that organization, meeting at Liberty, extended its full accreditation to the college while clearly "specifying a fully creationist position as a criterion for membership." During the high tide of creationist publicity in early

The National Creationist Movement

1982, Falwell said that his college's graduates would "go out into the classrooms teaching creationism. Of course they'll be teaching evolution, but teaching why it's invalid and why it's foolish." (*News & Observer* [Raleigh, N.C.], 30 May 1982; Morris and Rohrer 1982:229; Falwell 1981; Morris 1984:304–5; *Acts & Facts* [monthly newsletter of ICR], November 1983; author's notes on Moral Majority rally in Raleigh, N.C., 18 September 1981).

The sectarian connection that brings together Henry Morris, the Institute for Creation Research, Tim LaHaye, Christian Heritage College, Scott Memorial Baptist Church, Jerry Falwell, Liberty University, and Thomas Road Baptist Church is that all are Independent Baptist. This is a configuration of fundamentalists with Southern Baptist backgrounds who feel that the Southern Baptist Convention is insufficiently conservative and so have formed a breakaway faction to the right of the main Baptist denomination. These people and their institutions do not constitute a formal denomination but, rather, a loose network of like-minded conservative Baptists supporting each other's ministries. The Baptist preferences of Morris and his ICR associates are well reflected in the venues of their speeches, debates, and preaching. From October 1980 through October 1984, Baptist churches and other Baptist organizations accounted for 29.9 percent of ICR appearances, more than any other kind of church or organization (Toumey 1990a:120–121).

In truth, the New Religious Right is a diffuse, unstable realm, a shifting archipelago of autonomous churches, breakaway congregations, and local Bible chapels. No single denomination controls this loose alliance, but the Independent Baptist affiliation is the best sectarian network for getting in contact with the largest number of its sovereignties.

The moral theme of ICR's message to its constituents is that evolution contributes to spiritual decay in the United States by abetting Secular Humanism. LaHaye writes that "Evolution provides a theory of man's origin independent of the God the atheist believes does not exist. The atheist's next conclusion is, we believe, very dangerous to society. If man is an animal, then, like other animals, he is amoral" (LaHaye 1983:3).

Likewise, Morris says that "evolutionary humanism in our

56

schools is not only a religion, but is a religion which opposes Judaism, Christianity and the Bible in no uncertain terms" (Morris 1977). Furthermore, it is supposed that this evil influence can be countered by asserting that the biblical account of creation in Gen. 1:1–2:3 is literally true, and true in the sense that God has established an unchanging framework for conservative morality.

The fundamentalist Christianity of LaHaye, Morris, and Falwell must also come to terms with a second religion that coexists with the nation's various Judeo-Christian creeds: the religion of science and technology, represented by the trivial model of science in which nothing, not even a basic Bible story, is credible until it is surrounded by technicians, technical equipment, technological jargon, and technocratic authority. This is where the Institute for Creation Research excels. It gives conservative Christians the creation stories they want to hear with the moral meanings they require, and it sets them upon a stage of scientific sanctification decorated with test tubes, Kuhnian paradigms, white lab coats, monographs, geological expeditions, quotes from Karl Popper, and secular credentials. In other words, ICR performs the liturgies of popular science just as seriously as it executes the liturgies of conservative Christianity, implicitly accepting science as a credible form of knowledge as authoritative as the Holy Bible. This way it earns a double-barreled respectability from its followers, who need both moral values and scientific sanctification for those values. ICR assures them that the Genesis account of origins has been twice proven, once in the pages of Holy Scripture, where it is true because the word of God says so, and a second time, independently, in the abracadabra of scientific experts. Thus, scientific creationism, which insists that science is the Bible's best friend.

Even with this web of ideological sympathies and intimate connections, the relation between creationism and the New Religious Right is sometimes problematic, because creationism must compete with other issues and priorities for the attention and resources of sponsors like Falwell. Wholehearted commitment to scientific creationism at Liberty University flourished in the late 1970s and early 1980s, but Falwell's statements and policies brought trouble for the college. When students graduating in science education sought teaching jobs in the local public schools, the Virginia chapter of the American Civil Liberties Union challenged their qualifications, alleging that their knowledge was religious indoctrination rather than science education. To substantiate its case, the ACLU quoted Falwell's remarks. During a contentious evaluation by the Virginia Board of Education, the college recanted its explicit commitment to creationism as science. A. Pierre Guillerman, president of the school, diluted Falwell's earlier policy by saying that the school does not require students to accept creationism: graduation from Liberty, he said, "is in no way conditioned upon opposition to evolution" (Bentley 1984; *Christianity Today*, 3 September 1982:88–89).

In 1984 Liberty University established a Center for Creation Studies (CCS) to coordinate its teaching and research on creationism. For three days in April 1985 I visited Liberty University to learn the status of creationism there. Robert Littlejohn, professor of biology, and Lila Robinson, professor of anthropology, both stated that the college considered creationism a philosophical matter, not a scientific subject. Littlejohn confirmed that the college used no creationist texts in biology. Indeed, the bookstore had several creationist books, but none of the textbooks for biology or anthropology courses were creationist texts. The college catalogue for 1985 mentioned nothing about creationism, only that a student should have "a God-appreciation perspective by studying His revelation through creation, history, social processes, and the rational ability of man" (Liberty Baptist College 1985:7).

Lane Lester, director of the Center for Creation Studies, explained to me in an interview on 12 April that the school wanted to protect the job prospects of its science education graduates by avoiding another ACLU challenge. He said that Falwell suggested the CCS as a way to take creationism out of the science curriculum. "It was his [Falwell's] baby," Lester told me. He said that CCS was not to be connected with any science department. Instead, it would offer a general course on creationism, to be required of all students. Yet, as Professors Littlejohn and Robinson had said, it would treat the topic as a broad philosophical concern instead of representing it as scientific knowledge.

58

Not surprisingly, the doctrinal commitment endured despite this change of scientific status. On 13 April 1985, Falwell told an audience of prospective students at Liberty that if they came to the university, "You'll learn all about evolution, but you'll learn why you don't believe it. . . . To our knowledge, we've never graduated an evolutionist." Further, he said, "We don't have a faculty member here who doesn't believe the Bible is the inerrrant word of God" (author's notes).

Liberty University has institutional interests that demand difficult decisions about sectarian commitments. Falwell and Guillerman want the school to have the credibility of a major university, and they require the formal accreditation that makes their graduates employable. In these circumstances, the teaching of creationism as science is a liability to the institution. Isolating creationism within the Center for Creation Studies eliminates that liability. (It also gives Liberty the option to teach pure doctrine as such, without having to frame it as scientific hypotheses.) Thus Liberty has constricted its creationist connections to protect its institutional interests.

The leaders of the national creationist movement might well see Liberty's decision as a betrayal, since it is urgent to them to continue presenting creationism as science. They can take it for granted that a fundamentalist college like Liberty University will present it as Christian doctrine. That is not in dispute. The more urgent matter is whether the college will also give creationism an image of scientific authority by treating it as scientific knowledge. For Liberty to separate creationism from hard science must seem to the leading creationists to be a prevarication (which, in fact, it is, for it obscures the definition of creationism so that the ACLU cannot object).

Falwell and other famous preachers loudly proclaimed their creationist sentiments while creationism was a headline story, from spring 1981 through summer 1982. When that issue receded from the headlines, the luminaries of the New Religious Right invested their resources in other concerns, especially abortion, prayer in the public schools, and the reelection of President Reagan. Their doctrinal belief in creationism is no doubt sincere, but their political investment in it is obviously situational, for they must consider how it affects their own ministries and how it matches up with primary issues like abortion and school prayer. By these criteria, creationism is very much dependent on considerations beyond the control of creation-science believers.

Without the churches, academies, colleges, lobbies, rallies, broadcasts, and mailing lists of the New Religious Right, creationism would be an obscure oddity; with its Religious Right sponsors, however, it becomes a popular sensation that generates difficult public controversy about science education. But the price it must pay for its fame is to have its powerful Religious Right sponsors, who have bigger concerns, assign it to a secondary place in their order of priorities, leaving creationism vulnerable to its patrons' changing interests.

The Inerrancy Baptists

The Southern Baptist Convention (SBC), embracing fourteen million adult members, is the largest Protestant group in the United States. Its theology is simple: Southern Baptists agree on only two maxims, namely, that the Bible is the word of God, and that every Baptist has a solemn personal responsibility to study the Bible directly. But because different people read different things into the Holy Bible, they disagree about the meaning of the word of God. Also, because their different meanings are so personal, their many disagreements are intense. The *Christian Century* put it like this: "You know you're at a Southern Baptist Convention when . . . one pastor says 'The Bible is the authoritative and trustworthy word of God' and another pastor says 'the Bible is the inerrant and infallible word of God,' and that means the two of them are on opposite sides of the theological and political fence" (1–8 July 1981:694).

To prevent concentrations of personal power, SBC limits its presidents to two consecutive one-year terms. Consequently, Southern Baptists have an occasion for confrontation with each other at least every other year. Baptist theology and Baptist politics thus govern the way SBC encounters the question of creationism. First, it is reduced to a dispute over the meaning of biblical authority, usually dividing Baptists into a camp favoring the narrowest possible inerrancy, and a camp with a slightly more relaxed exegesis. Secondly, whatever is resolved in one year can be argued again in another year or two. Old Baptist issues never die. Instead, the factions within the SBC replace their representatives on a regular basis, with the consequence that each new round of leaders must establish its doctrinal credentials de novo by resurrecting old issues.

Often they choose creationism as the best idiom for announcing their positions on scriptural authority. In the early 1980s, inerrantists within SBC raised accusations that (1) a religion textbook used at Baylor University in Waco, Texas, was unacceptable on the grounds that it "presents Adam and Eve as symbols rather than as historical persons"; (2) an unnamed professor of religion at Baylor had referred to the two creation stories in Genesis as political rhetoric rather than historical fact; (3) a Baylor professor (again unnamed) believed evolution was a part of God's plan for creation; and (4) faculty at Southeastern Baptist Theological Seminary in Wake Forest, North Carolina, ought to be "folks who have no questions about the full validity of the scriptures" (*Christian Century*, 10–17 September 1980:839–840; *Biblical Recorder* [Cary, N.C.], 14 July, 31 March 1984).

Ever since the inerrantist faction took control of the presidency of SBC in 1979, its candidates have frequently invoked creationism to establish their ideological credentials. SBC president Rev. Bailey Smith of Del City, Oklahoma, proclaimed in 1980, "I believe God created the world in six 24-hour days"; when he left office in 1982, he stated that it is inexcusable "for a Southern Baptist to say Genesis is political rhetoric and not historical fact [or] . . . for a Southern Baptist to teach evolution in our schools" (*Charlotte Observer* [Charlotte, N.C.], 13 June 1980; *News & Observer*, 16 June 1982). His successor, Rev. James T. Draper, Jr., reaffirmed creationism as inerrancy: "I'm not going to pay the salary of someone who said Adam and Eve were fictitious . . . . If that event is fictitious, then there wasn't a fall. The story of creation is essential, a foundation . . . Evolution is a theory. There is not a shred of evidence showing the existence of vertical evolution" (*News & Observer*, 17 June 1982).

The next president of SBC, Rev. Charles Stanley of

Atlanta, Georgia, said in 1985, "If you've got people here [at a Baptist seminary] who don't believe the first eleven chapters of Genesis, you need people who do" (*News & Observer*, 4 April 1985).

Creationism as a point of contention in these Baptist arguments ordinarily appears only in the context of inerrancy, unburdened by the scientific connotations preferred by Henry Morris and ICR. And although the Southern Baptists occasionally lump the matter of creation versus evolution with other social conflicts, they usually dwell on narrow battles of exegesis and exegetical heresy, thereby relieving the issue of creationism of the New Religious Right's sweeping social theory that ties evolution to an array of evil and immorality. Though the boundaries are sometimes blurred, those considerations make the inerrancy Baptists' moral interpretation of creationism and evolution substantively different from that of the New Religious Right.

The Evangelical Center

The evangelical center is the third major U.S. Protestant group for whom creationism is a serious ideological question. The group's problem is especially poignant: creationism seems to be sound religious truth by virtue of its biblical source, but it can subvert the mission of bringing souls to salvation because of its sharply controversial image.

Evangel comes from a Greek expression meaning good news, in this case, the gospel of salvation through Jesus Christ. An evangelist, whether preacher or layperson, is a messenger bringing the Good News. This definition presupposes some kind of belief in biblical authority, but most of the Bible is dim background for the bright message of Christian salvation. Many evangelicals are content to reduce their belief to the twenty-five words of John 3:16: "For God so loved the world, that he gave His only begotten Son, that whosoever believeth in him should not perish, but have everlasting life." As a result, evangelical theology is often superficial.

Furthermore, the Good News has a totally inclusive meaning, theoretically capable of moving and saving everyone who hears it. Issues of exegesis and inerrancy, by contrast, are notoriously divisive, with people disagreeing bitterly among themselves about how to interpret scriptural authority. Evangelical Christians know quite well that news that comes from a messenger with a chip on his or her shoulder is not received as good news. Accordingly, their doctrine is vague, which renders it unavailable as ammunition in the schisms that threaten the work of converting the world.

From these considerations, some important distinctions can be drawn between the evangelicals and their fundamentalist cousins. The latter care principally about conforming to conservative standards, whether the social morality of the New Religious Right or the biblical exegesis of the inerrancy Baptists. Fundamentalists gauge purity in degrees of schism, while evangelicals measure success in conversion statistics. Evangelicals will likely focus softly on the four Gospels and a handful of Epistles.

These idealized differences do not always produce clear distinctions between these kinds of Christians. Many evangelicals are moderately conservative, if not fundamentalist; most fundamentalists consider themselves evangelicals; several million U.S. Protestants occupy the overlap of the two categories; many are morally fundamentalist in their private lives, and pragmatically evangelical in their public lives.

Nevertheless, there are some recognizable evangelical stances and institutions. Regarding creationism and evolution, the American Scientific Affiliation is the most important organization of evangelicals. It was formed in 1941 by evangelical Christians in the scientific community as a remedy for the ignorance that plagued creationist sentiment (Numbers 1982). This was not in any sense an anticreationist organization but, rather, a group that hoped to relieve Protestantism of the more embarrassing features of extreme creationism. The Deluge Geology Society, a group of old-time Seventh-day Adventist creationists, coexisted with ASA for a few years, but after the society's demise in about 1945, most hard-line creationists merged into ASA because they had no place else to go. Later, when the hard-liners under Henry Morris reformulated biblical creationism in the 1950s and 1960s, ASA maintained a flexible, moderate position on origins. The American Scientific Affiliation asserted, in effect, that the natural world was a creation of God, but that it was also acceptable for a Christian to believe that evolution is part of God's plan.

As a result of their religious concerns, ASA's three thousand members shun the more atheistic or mechanical versions of evolutionary thought, but as a result of their scientific judgment, they doubt many features of Henry Morris's version of creationism. Most reject his young-earth chronology, which sets the age of the earth at ten thousand years or less, and his flood geology, which supposes that Noah's Flood was responsible for most of the geological evidence we find today.

Even more significantly, ASA members challenge hardline creationists on religious grounds. In *Origins and Change*, an ASA reader on the creation-evolution controversy, Richard H. Bube, a professor of materials science at Stanford, alleges that "to pose such a choice [namely, that creation and evolution are mutually exclusive,] is to do basic damage to the Christian position. It is to play directly into the hands of those evolutionists who argue that their understanding of evolution does away with the theological significance of Creation . . . The Christian anti-evolutionist is wrong to believe that his theological description must make any biological description impossible" (Bube 1978:vii).

Statements like this from ASA and its members are set within a carefully balanced context that proclaims creation as a theological truth that should not constrain scientific thought, and that accepts evolution as an empirical reality with no religious significance. As Bube writes in the same article, evolution and creation are "descriptions of the same phenomena on different levels of reality . . . Evolution *can* be considered without denying creation; creation *can* be accepted without excluding evolution. Evolution is a scientific question on the biological level; it would be unfortunate indeed if a scientific question were permitted to become a crucial point for Christian faith" (1978:vii).

The fundamentalist creationists have no sympathy for ASA's intellectual balance. In their eyes, ASA's policy is the worst kind of character weakness. As Christians, the members of the American Scientific Affiliation should acknowledge the

64

biblical truth of the creationist cause, but because they care too much for the approval of their fellow scientists, they betray the cause. The materialistic evolutionists are easier to understand. They are depraved because they are atheists. The members of ASA are held to a different standard: Christians should know better and should close ranks with the hard-core creationists.

Henry Morris blasts ASA in a long diatribe that dwells on its character weaknesses. ASA, he charges, has "capitulated to theistic evolution." It has led Christians into compromise; it is guilty of backsliding and of drifting toward "total evolution" (Morris 1984:130–142).

Wheaton College in Wheaton, Illinois, where many ASA members have been students or faculty, draws the same kind of hostility from Morris and the Institute for Creation Research. They complain that the college is intolerably ambiguous in its position on creationism, and that this is so because of a desire for academic respectability (Morris and Gish 1976:79–82). In his *History of Modern Creationism*, Henry Morris denounces certain Wheaton faculty for contradicting his creation chronology, and he accuses them of "persuading thousands of evangelicals to follow down the fatal path of compromise with evolution" (Morris 1984:113–116).

The magazine *Christianity Today* reflects the evangelicals' acute ambivalence, as it vacillates between favoring creationism subtly but never quite endorsing it whole-heartedly, and almost disowning creationism without really condemning it. This magazine addressed the creation-evolution dispute in June 1977 with two articles and an editorial. The editorial, by Harold Lindsell, had a creationist tone, denouncing naturalism, uniformitarianism, and materialism (Lindsell 1977). Lindsell endorsed creationism for theological reasons, but he avoided discussing its scientific merits. An accompanying article by Tom Bethell blasted Darwinism on scientific and philosophical grounds, charging facetiously that "survival of the fittest" is a tautology fatal to Darwinism, that Darwin "is in the process of being discarded," and that "it is being done as discreetly and gently as possible, with a minimum of publicity" (Bethell 1977). Though rejecting Darwinism, Bethell did not embrace creationism.

The third piece in that issue of *Christianity Today* was an

65

interview with Jack Haas and Richard Wright, two scientists in the American Scientific Affiliation (17 June 1977:8–11). Its substance was a sharp denunciation of scientific creationism: the evolution model was the only scientific model of origins, and most Christian colleges were comfortable with theistic evolution as a philosophical framework. The editorial and the two articles each had a firm view of creationism or of evolution, but taken together as a whole that might reveal the magazine's stance on scientific creationism, they had the effect of talking around it without taking any firm stand.

Three years later, *Christianity Today* referred offhandedly to "the so-called theory of scientific creationism," but within another year it called evolution "science's sacred cow," and complained that evolutionists were "not willing to face creationism on its merits" (18 April 1980; 20 February 1981). That was in 1981, when creationism was an exciting Christian message. But in 1982, after legal disaster in Arkansas and popular ridicule everywhere had made creationism a liability for Christian evangelism, *Christianity Today* published a vehement denunciation of creationism by Edwin A. Olsen, a geology professor at a conservative Presbyterian college in Spokane, Washington (Olsen 1982). Olsen said that creationists were intolerant, simplistic, and less honest than evolutionists about their own failings. He cut right to the heart of creationism's appeal to evangelical Christians: "I consider the creationist approach to be poor strategy for a truly Christian impact on the world." Further, "creationism . . . is a caricature of the true Christian view of men and things. In its isolation and inflexibility, 'creationism,' in my judgment, is doing more harm than good."

Olsen's article was shockingly blunt and, in retrospect, was more important for evangelical Christians than any other piece in *Christianity Today*. More than the arguments about thermodynamics, stratigraphy, or the statistics of randomness, Olsen addressed the question that evangelicals cared most about: What does creationism do for the Good News of salvation through Jesus Christ? In effect, Olsen answered that creationism perverts the Good News.

In its next issue, *Christianity Today* dissociated itself from Olsen's views, but three months later it printed a friendly interview with the astronomer Robert Jastrow, who reasserted his firm belief in evolution (7 May:12–13, 6 August 1982:14–18). The last major statements on creation and evolution appeared in October 1982. Duane T. Gish and Thomas G. Barnes of the Institute for Creation Research restated the creationist orthodoxy, while V. Elving Anderson and Davis A. Young presented the American Scientific Association's consensus, that evolution is an empirical reality whose moral meaning must be set within a framework of Christian values. The editors let the controversy end in a standoff between ICR and ASA (*Christianity Today*, 8 October 1982:28–45).

Christianity Today, with its oscillations, vacillations, and qualifications, is an apt microcosm of evangelical Christians' sentiments. Many evangelicals sympathize with the creationist cause because of its biblical inspiration, but when creationism becomes divisive in its tactics or disreputable, they must separate their Christian ministries from creationism lest it jeopardize their chance to convert more souls to Jesus Christ. Yet rejecting creationism must not lead one to embrace evolution, or at least the atheistic versions of evolutionary thought. Evangelical statements on evolution are often overly precise, making exquisitely subtle distinctions among numerous definitions of evolution according to which ones include God explicitly, and how; which ones allow God implicitly; which ones exclude God; which ones make no connection either way; and so on.

Most evangelical Christians in the United States, as individuals, come to their own reasonable conclusions about how evolution, creation, and Christianity fit together. Yet they are chronically unable to formulate group positions, institutional policies, or corporate principles for this controversy. Their public positions, as in the case of *Christianity Today*, are usually convoluted or contradictory. The evangelical center cannot embrace scientific creationism, but it cannot renounce it, either.

The Apocalyptic Separatists

Another stream of creationism runs parallel to but separate from the scientific creationist movement. This is the anti-evolutionism of Jehovah's Witnesses and the Worldwide Church of God.

The Witnesses proselytize door-to-door, making their magazines, *Awake!* and *Watchtower*, widely available, while the Worldwide Church of God distributes its magazine, *Plain Truth*, in dozens of airports and thousands of shopping centers. Despite their extensive proselytizing, however, these two sects have little influence outside their own respective circles of believers, largely because each is intensely hostile to most features of mainstream U.S. culture, including both Protestant orthodoxy and positive regard for science. Each has a long tradition of opposing evolution, but both are careful to isolate themselves from other churches, from other creationists, and from each other.

Three general features characterize apocalyptic separatism: (1) Cosmic history focuses on a titanic primal stuggle between God and Satan, soon to be concluded in an Apocalypse, in which humans are quite insignificant; (2) the outside world, including all other churches, is believed to be a dangerous land infested with satanic tricks, giving these two sects reason to stay apart from all others; and (3) their respective internal politics require unwavering obedience to theocratic authority, leading them to define the problem of evolution as subversion of obedience.

The Worldwide Church of God considers the entire outside world, including all other churches, to be "mentally, morally, and spiritually sick, filled with headache, frustration, broken homes, dope, crime, insanity, perversion, racial hatred, poverty, filth, pollution, disease, and traffic jams" (Martin 1973), and the Witnesses have a similarly vehement sense of separateness (Beckford 1975:104). In the words of one Witness I interviewed in North Carolina: "We don't believe in interfaith. We don't identify with other religious groups. We have no coalitions with them. . . . There are groups that are trying to change the world. We believe that only God's plan, as revealed in the Bible, can save the world. . . . Falwell and those [others] wave the American flag; it's more important to them than the religious aspect."

The Worldwide Church of God has enjoyed an unbroken tradition of intense anti-evolutionism since January 1934, when Herbert W. Armstrong founded that sect, and even after his death in January 1986. Armstrong decided in 1927 that rejecting evolution was one of his original "seven conclusions"

68

that would guide his life and his church. "Evolution," he wrote back then, "could not honestly be reconciled with the first chapter of Genesis!" Therefore, "evolution stands disproved—an error—a false theory" (Hopkins 1974:30–31; Armstrong 1984).

The Jehovah's Witnesses and the Worldwide Church of God think of their respective beliefs as being quite distinct from those of all other religious groups, but some of the standard arguments advanced by the Institute for Creation Research creep into their publications. For eighteen years, the Witnesses' principal reference work on the creation-evolution controversy was a little blue book published in 1967 that contained much that the Witnesses' writers derived from Henry Morris and his friends (Watchtower BTS 1967). Not until 1985 did they publish a new book that eliminated their explicit reliance on Morris and ICR (Watchtower BTS 1985). Their basic proselytizing pamphlet on anti-evolutionism, a special edition of *Awake!*, concurred with the ICR line in all technical matters except chronology (Watchtower BTS 1981). Likewise, the Worldwide Church of God has included much mainline creationism in its own works. Articles in its monthly magazine, *Plain Truth*, and in its many topical pamphlets agree with ICR's positions on numerous issues (Steep 1981, 1982; Worldwide Church of God 1971).

Those concurrences do not mean that the two sects agree with the theology of the scientific creationist movement. They reveal rather that Jehovah's Witnesses and the Worldwide Chruch of God are unable to generate original sceintific literature. Neither sect has either scientific staffs or scientific institutions. Ambassador College, a branch of the Worldwide Church of God in Pasadena, California, had only two M.S. degrees, one B.S. degree, and two Ph.D.'s (in psychology) among the credentials of its entire faculty, according to the college's 1983–85 catalogue. (Some part-time faculty members had Ambassador College degrees.) Jehovah's Witnesses have no college of their own, let alone any science departments. It is inevitable, then, that these two groups would borrow scientific information from groups they would otherwise prefer to avoid (Toumey 1987:235, 1990a:133–139).

For groups that cherish separatism as intensely as these two, it is embarrassing to have to borrow information about creation. They need some distinctive symbols or idioms to re-establish their respective postures of independence. One of their ways of doing this is to outflank the scientific creationist movement on the right by throwing the issue of scientific credibility back in its face, accusing the main creationist groups of being too scientific. The apocalyptic separatists distinguish between *creation*, which to them is the incontrovertible biblical teaching about origins, and *scientific creationism*, by which they mean the irresponsible habit of diluting biblical truth to satisfy scientists by equating it with scientific theory. Thus, Sidney M. Hegvold of *Plain Truth* writes that, "since scientific theories are continually changing and cannot provide absolute truth, should creation be considered and treated as just another scientific *theory?* . . . Clearly it cannot! . . . It would be a greater tragedy [than the teaching of evolution] if now . . . creationists should be given equal foothold to treat their personal ideas of creation as a nonreligious scientific theory!" (Hegvold 1982 [original emphases]).

Here Hegvold has caught the scientific creationists red-handed, for indeed their principal claim is that, for purposes of public school education, God's creation can be transformed into a religiously neutral scientific model. The genius of this denunciation is that the separatists are doing to the scientific creationists just what the creationists are doing to evolutionists; namely, accusing them of being so seduced by the values of the secular world that they refuse to accept simple biblical truth. Any kind of participation in the secular world can be used as evidence that the accused has been seduced by secular values. With this moral one-upmanship, the apocalyptic separatists reestablish their apartness from the scientific creationist movement. This stance is also illustrated in the titles of *Plain Truth* articles like "Evolutionists and Creationists Are at It Again!" (Elliot 1983) and such *Awake!* articles as "Evolution, Creation, or Creationism: Which Do You Believe?" (Watchtower BTS 1983b).

Another matter that separates these groups from the scientific creationists is biblical chronology. The Worldwide Church of God subscribes to a gap theory that posits an enormous length of time after the initial creation in Gen. 1:1 but before the six-day creation that starts at Gen. 1:3 (Steep 1983). During this time, they believe, Satan struggled against God, leaving the earth as the detritus of their battle. With this view, long geological ages can represent that long period of struggle, and violent geological events like earthquakes and volcanic eruptions can reflect the violence that God and Satan inflicted on each other. Although no geologist takes it seriously, this theology uses geological evidence as a foil to Henry Morris's flood geology.

The Jehovah's Witnesses also invoke geology to refute flood geology (Watchtower BTS 1983a, 1983b). They combine an unconventional biblical exegesis with an unusual interpretation of Archbishop Ussher's reference date for creation (approximately six thousand years ago) to produce a chronology in which the days of creation were not literal twenty-four-hour days (compare Ps. 90:4 and 2 Pet. 3:8) but units equal to seven thousand human years. Thus the six days of creation in Gen. 1 lasted a total of forty-two thousand of our years. Jehovah's Witnesses then fix the end of the creation process—the end of the sixth day—by locating the creation of Adam and Eve at 4025 B.C., which is a minor variation on Ussher's chronology. This chronology is so esoteric that there is no danger of its being co-opted by any other religious group or creationist institution. Its mathematical particulars are not especially important to the internal relationships of this sect, but they are quite useful as an external barrier between the Witnesses and the outside world, including the scientific creationist movement.

The Worldwide Church of God and Jehovah's Witnesses each have intensely authoritarian structures that require unqualified obedience. For both, religion, including creationism, is essentially a business of conforming to church authority. The problem of evolution, in their view, is that it subverts obedience to the Creator and the church authorities (Watchtower BTS 1967; Whalen 1962:33–38). In their commentaries on evolution, they emphasize the tragedy of disobedience. The Worldwide Church of God even generalizes its fear of evolution to apply to all scientific thought. The scientific mind is said to be a rebellious mind. Armstrong set the pace of the church's hostility to science with a steady stream of denunciations prefaced by, "scientists and educators believe that . . . ," or "scientists and educators teach that . . .":

> To replace religion and belief in God, scientists and educators had substituted the doctrine of evolution. The tools of modern science used in the production of this new KNOWLEDGE were a stepped-up use of those man had employed since the dawn of history— rejection of revelation as a source of knowledge and the use of observation, experimentation, and human reason. (Armstrong 1983a [original emphases])

> Has either science or education PROVED the evolutionary theory? Emphatically they have not! . . . Has either science or education PROVED the *non*-existence of a personal Supreme God? . . . Emphatically they have not! Why, then, do so many great minds who profess knowledge and wisdom doubt or deny God's existence? Simply because of something inherent in human nature—something of which they are ignorant—a spirit of vanity, coupled with hostility and rebellion against their Maker and His authority! (Armstrong 1983b [original emphases])

Plain Truth elaborates on this theme in articles titled "What Spokesmen for Science Are Not Telling" (Stenger 1984) and "What's Wrong with Science?" (Elliott 1982). This second article presents the church's disposition most clearly by equating science with the original sin: "Just as Adam and Eve, scientists have rejected instructions from the Creator. . . . The scientific world has the attitude that motivated man's first sin!" The author then predicts the consequence of the scientific attitude: "The Bible says that a day is coming when God will again intervene and save this world from destroying itself. Then science will be put in its proper place. . . . Mouths will drop open, knees will shake" (Elliott 1982).

While denouncing scientific values, the Worldwide Church of God also employs scientific knowledge to support its claims of godly authority. The usual pattern of a *Plain Truth* article on a scientific topic such as evolution is to begin with a factual and interesting introduction, using reliable sources like encyclopedias to establish the magazine's credibility. After winning the reader's interest, a transition of two or three paragraphs shifts to the theme of confusion and uncertainty, pointing out the limits of human knowledge, emphasizing that the supposed scientific experts are not as smart as they claim to be. Next, a series of rhetorical questions elicits the reader's anxiety regarding the dissolution of scientific expertise, evoking a desire for an authoritative answer: "Why don't the experts admit that . . .?"; "Is it true that . . .?"; "What would happen if . . .?" Finally, the article solves the problem it has constructed by stating that Herbert W. Armstrong deciphered biblical passages pertaining to the problem, and that the church is willing to share his understanding with the reader. The Worldwide Church of God never admits to a Baptist-style biblical exegesis, that is, the view that ordinary individuals can understand biblical truth just by reading the Bible by themselves. Instead, it insists that its founder, Armstrong, possessed a special ability to decode the mysteries of the Bible, making church members dependent on his inspiration and subservient to his authority.

Although they blast evolutionary thought, the Apocalyptic Separatists seldom single out individual evolutionists for personal blame or ad hominem attacks. This style is unlike that of the scientific creationists, who often berate Stephen Jay Gould, Carl Sagan, the late Isaac Asimov, and other scientific celebrities. The separatists' enemy is evolution as an evil idea, not evolutionists as agents of evil. This is because their view on moral causation revolves around the cosmic struggle between God and Satan. The critical events of the apocalyptic chronicle happened long before humans were created—hence the need for gap theory to allow time for those events prior to the creation of Adam and Eve—and the next truly important events will happen at the Apocalypse, when God and Satan conclude their violence against each other. Human history is merely an interlude between the really important times. Humans, whether evolutionists or creationists, do not influence this kind of history; all they can do is line up obediently on the side of God or the side of Satan. As a result, evolution and other evils are seen to be more the works of Satan than of sinful humans. The Jehovah's Witnesses' blue book on evolution states that "responsibility for worldwide wickedness, then, rests primarily with Satan the Devil" (Watchtower BTS 1967), and *Plain Truth* declares, "What we call 'Human nature' is actually Satan's nature!" (Stump 1981).

The Evils of Evolution

Each of these factions shapes creationism to its own needs, deciding how much concern to invest in this issue and which moral meanings to draw from it. The New Religious Right, especially its Independent Baptist component, looks to the Institute for Creation Research to affirm that the creation-evolution controversy is a part of a grand moral struggle of Secular Humanism against conservative Christianity. This means that evolution is not necessarily the direct cause of moral decay: sometimes it is seen as a symptom of Secular Humanism and sometimes as a source, but in either analysis Secular Humanism displaces evolution as the principal problem facing Bible-believing Christians. If they can solve the primary problem of humanism, they feel, the secondary problem of evolution will become that much more tractable. Furthermore, the Institute for Creation Research has an irreducible capability to cast its creationist views in the popular idioms of scientific credibility, thereby sustaining so many Christians whose faith is less than secure unless it has a semblance of scientific sanctification.

The inerrancy Baptists, pressing their concerns within the Southern Baptist Convention, certainly share many sentiments with the New Religious Right, but they turn their attention to a special priority—biblical inerrancy. This, for them, is the most important issue (much more important than the question of whether science concurs with scripture). Although these Baptists may not intend it, the business of inerrancy can become a self-contained idiom, shutting out the long laundry list of moral concerns that bother the New Religious Right.

Next to consider is the evangelical center and the American Scientific Affiliation that serves it. Because the evangelical center is so pragmatic about conversions, it is leery of white-

74

hot controversies like the creation-evolution dispute. ASA, reflecting this view, avoids taking much of a specific stand on that issue. Consequently, it is notable more for what it avoids than for what it endorses. It denounces atheistic versions of evolutionary thought, but it also decries extreme expressions of anti-evolutionary thought. As a Christian organization, ASA feels that one should appreciate the moral dimensions of God's creation, but as a scientific organization, it does not expect creation to be tangible in a scientific sense. Because ASA and the evangelical center are neither clearly creationist nor clearly evolutionist, their thoughtful reflections have little impact on the creation-evolution controversy.

Lastly, the apocalyptic separatists: sectarian differences are everything to both the Worldwide Church of God and Jehovah's Witnesses. Each has a cosmology that describes the world around it, including U.S. culture, as a despicable condition corrupted by the tricks of Satan, including the apostasies that corrupt all other churches. To these worried people who see themselves as isolated pockets of God's true believers, all of human reality, including science, is hardly worth knowing except as a cunning ruse that can seduce a person from the obedience to theocracy that can save one from the doom that awaits others. In this intensely hostile atmosphere, science can never be anything better than an enemy of creation. No danger here of mingling science and scripture: the apocalyptic separatists hate science so much, they feel the scientific creationists are traitors to the Holy Bible for incorporating scientific idioms into biblical exegesis.

A common philosophical thread runs through all the Christian forms of creationism: the more one accepts that Bible stories are real, particularly those in the first eleven chapters of Genesis, the stronger God's people are in their struggles against sinful forces. But this is a thin thread, and a bare one. It needs to be strengthened with lively moral meanings about humanism, science, inerrancy, conversion, Satan, and so on before creationism acquires a tangible fabric that Christians can appreciate. As it turns out, the various sectarian factions attach different meanings to their own parts of the common thread, caring little whether the separate sections connect with each other. The

75

inerrancy Baptists offer a version of creationism so concentrated on one theme, biblical authority, that it often ignores broader moral concerns. The evangelical center has a sense of creationism so diluted, lest it contaminate the work of conversions, that its members' sincere sympathy for the creationist cause is barely perceptible. For apocalyptic separatists, creationism is yet another device for segregating themselves from all others, including those in the scientific creationist movement.

Five
Evolution and Secular Humanism

The core of the New Religious Right's moral theory is the fear that a conspiracy named Secular Humanism causes the evil and immorality that permeates the modern world. Creationism is an exercise in applying that theory to the question of origins, by charging that the idea of evolution is intimately connected to Secular Humanism. Therefore, to understand this ideological basis of creationist thought, it is necessary first to understand what fundamentalists mean when they speak of Secular Humanism.

A Brief History of the Idea of Secular Humanism

For most of U.S. history, Protestant culture has dominated public life, and especially the cultural climate of public schools. Evangelical prayers, Bible devotionals, the Common Sense philosophy, admonitions (justified by scriptural proof-texts) to shun the common vices, the evangelical ethos of proselytizing with one's personal witness, the piety of the born again, traditional sex roles: these features constituted much of the fabric of normative values. Religious minorities—Catholics, Mormons, Amish, Jews, Jehovah's Witnesses, and others—usually conceded the mainstream culture to the Protestant style and withdrew into unique religious subcultures to privately exercise their own faiths. Whether or not most people observed the Protestant code in their own behavior, it still dominated public culture. It is not hard to see why many Protestants sincerely believed that the United States was an intrinsically Protestant nation.

77

The National Creationist Movement

By the late 1950s, however, the Protestant hegemony was coming unglued. Consider public norms for sexual morality. In 1956, a U.S. Senate subcommittee urged that young people needed better sex education, and more of it (*New York Times*, 21 May 1956), while the United Lutheran Church heard its commission on family morality recommend that the church relax its laws on divorce and birth control (*Newsweek*, 2 January 1956). The following year, FBI director J. Edgar Hoover advised the nation that an increase in pornography had caused a rise in sex crime (*New York Times*, 3 May 1957). The Kinsey Institute reported in 1958 that 10 percent of "upper-class" women became pregnant before marriage (*New York Times*, 25 February 1958), and in 1959 a high school teacher in Van Nuys, California, sparked a scandal by eliciting a survey of his students' sexual experiences (*New York Times*, 9 August 1959). *Playboy* made sex almost respectable at the newsstand. The birth control pill, forever to change sexual choices, was approved for public use in May 1960. In short, sex roles and sexual morality, along with the Protestant values that defined them, seemed to many people to be under siege.

A series of developments in constitutional law further eroded the Protestant hegemony, as religious minorities established their rights to participate fully in public life and the public schools, without having Protestantism forced on them. Through the 1930s, 1940s, and 1950s, the U.S. Supreme Court used the Establishment Clause of the First Amendment to fix a new balance between non-Protestants and the old Protestant hegemony by diminishing the legal status of the latter. The plaintiffs in almost all of these skirmishes were Catholics, Jehovah's Witnesses, Seventh-day Adventists, Unitarians, and other sincerely religious parties. Few were atheists or agnostics. Regardless, some Protestants interpreted these events as an attack on U.S. culture by well-organized enemies of religion.

The suspicion that an evil conspiracy of unbelief had caused these changes was fueled by the two landmark freedom of religion cases of the early 1960s, namely, *Engel v. Vitale* of 1962 (370 US 421) and *Abington v. Schempp* of 1963 (374 US 203), in which the U.S. Supreme Court ruled that public schools must not force either group prayer or Bible devotionals, respectively,

78

Evolution and Secular Humanism

on their students. From the text of a third decision, *Torcaso v. Watkins* of 1961, came a name for the supposed conspiracy: "Secular Humanism." The plaintiff Torcaso, an atheist, had been denied the office of notary public by the state of Maryland because he would not affirm a belief in God. The U.S. Supreme Court held that he was entitled to that office, since Maryland's policy violated Article VI of the U.S. Constitution ("No religious test shall ever be required as a qualification to any office or public trust under the United States"). In a minor footnote to his opinion, Justice Hugo Black commented that "among the religions in this country which do not teach what would generally be considered a belief in the existence of God are Buddhism, Taoism, Ethical Culture, Secular Humanism, and others" (367 US 495). The Court did not define that last term, however, nor did it give Secular Humanism any special attention. If any formal status for Secular Humanism could have been inferred from *Torcaso*, it was no more significant than, say, the status of Taoism.

Four years later, in the *U.S. v. Seeger* decision of 1965, the Supreme Court concluded that a sincere personal belief in a Supreme Being constituted sufficient religious grounds for conscientious objector (c.o.) status for draftees. In doing so, both Justice Tom Clark, author of the majority opinion, and Justice William Douglas, writing a concurring opinion, mentioned that a professed atheist could not use the *Seeger* decision to achieve c.o. status, for obvious reasons. Justice Douglas added a brief footnote that cited *Torcaso* as an example of an atheist's beliefs (380 US 193).

Torcaso and *Seeger* came to be cited in fundamentalist folklore as a pair of decisions that made Secular Humanism an "official" U.S. religion. Examples are the writings of John Whitehead and John Conlan (1978), Tim LaHaye (1980), Onalee McGraw (1976), and the ProFamily Forum (1980). Apparently the reasoning of those authors was that if the Supreme Court merely mentioned it, that makes it official. This interpretation is especially ironic, since it takes the two footnotes more seriously than the primary texts they accompany, and because neither text offers any definition of the critical term. Furthermore, *Seeger* is worse than irrelevant to the issue of legiti-

79

mating Secular Humanism, since it expands the significance of theistic religion by giving special status to those who believe in God, while the comment by Justice Douglas plainly separates someone in the position of Torcaso, an atheist, from the benefits of this religious status. Nevertheless, *Torcaso* and *Seeger* are the origins of Secular Humanism's official standing, according to fundamentalist belief.

In *Abington v. Schempp*, the Supreme Court commented that "the State may not establish a 'religion of secularism' in the sense of affirmatively opposing or showing hostility to religion, thus 'preferring those who believe in no religion over those who do believe' " (374 US 225). This brings us to two competing theories about the legal status of Secular Humanism: did the Supreme Court establish a religion of Secular Humanism in *Torcaso* and *Seeger*, or did it disestablish such a religion in *Abington?* In fundamentalist belief, the answer is both. The establishment theory crystallized resentment against the loss of the Protestant hegemony, while the *dis*establishment theory inspired legal assaults against secular knowledge, including evolution and sex education. For example, in 1969 Max Rafferty, the ultraconservative superintendent of public instruction of California, produced a document that described a moral crisis in the United States and blamed it on Secular Humanism (Rafferty 1969). Sex education, behaviorism, Marxism, and evolution, Rafferty said, could be traced to Secular Humanism, which he described as "a philosophy of life which rejects traditional standards of 'morality' " (11). Citing *Torcaso*, he said that "humanism is, by definition, a religion" (42); citing *Abington*, he concluded that the California public schools must not teach that religion (71–72). "The need today," he stated, "is to contrast the American genius and the American's reliance on Almighty God with the cold, dreary utilitarianism of the Secular Humanists or Marxists" (68).

Three years later, in 1972, William Willoughby, religion editor of the *Washington Evening Star*, brought suit to compel the National Science Foundation to balance comments on evolution in its publications with statements giving equal credibility to creationism. Otherwise, he contended, the government would be "establishing secular humanism as the official religion of the United States" (Nelkin 1982:100–101). Willoughby's suit failed, both in federal district court and at the Supreme Court. In 1976, Representative John Conlan, a conservative Republican from Arizona, introduced an amendment to that year's education appropriations bill, stipulating that no federal funds could be expended in support of "any aspect of the religion of secular humanism." His amendment passed in the House but died in the House-Senate conference. Also that year, the school board of Frederick County, Maryland, prohibited "any persuasion of humanism that promotes a religious or irreligious belief" (McGraw 1976:8–9). Two years later, in 1978, Dale Crowley of the National Bible Knowledge Association sued the Smithsonian Institution on the grounds that an evolution exhibit constituted an establishment of the religion of Secular Humanism. His initiative failed in federal court (*New York Times*, 12 April, 14 December 1978, 19 May 1979; *Science*, 1 June 1979:925).

A more significant event of 1978 was the appearance of a law review article by John W. Whitehead and John Conlan, wherein the authors contended that Secular Humanism is a religion in First Amendment terms (Whitehead and Conlan 1978). Two features of that article are especially significant: it offered a theory of the history of Secular Humanism, and it gave some substantive content to that term. The historical theory began with the claim that colonial society was so intrinsically Protestant that Protestantism should still rule U.S. life. But, continued Whitehead and Conlan, Secular Humanism usurped that hegemony by employing a series of wrong-minded Supreme Court decisions, beginning in 1878 and culminating in the *Seeger* case of 1965. Currently, they asserted, Secular Humanism is so deeply embedded in public life and government policy that it occupies the status that only Protestantism deserves to own. Yet its very success makes it an "established" religion, in First Amendment terms. Thus Whitehead and Conlan offered fundamentalist leaders great comfort by predicting that the First Amendment could wreck their enemy as surely as it had wrecked public school prayer. (For a more conventional account of the same legal history, see Hammond 1984.)

Furthermore, the authors solved a serious problem in the argument against Secular Humanism by doing something the

Supreme Court had never done, namely, define the phrase. After studying a pair of documents by the American Humanist Association called Humanist Manifestos I and II (*New Humanist*, May–June 1933; *Humanist*, September–October 1973), Whitehead and Conlan stated that "Secular Humanism is a religion whose doctrine worships Man as the source of all knowledge and truth, whereas theism worships God as the source of all wisdom and truth" (Whitehead and Conlan 1978:30–31), and that "along with the evolutionary theory, the centrality and autonomy of Man are the prominent features of Secular Humanism" (44). From the comment on autonomy, they equated Hitler and Stalin with humanism (45), and elsewhere they labeled Secular Humanists as "those who believe in no morals" (19). Thus, after seventeen years of citing one sparse footnote from *Torcaso* and a second, even sparser, one from *Seeger*, fundamentalist Christians finally had something about Secular Humanism they could describe in detail.

Even at that point, hostility to Secular Humanism was an obscure legal theory. But in 1980 a book titled *The Battle for the Mind*, by Rev. Tim LaHaye of Scott Memorial Baptist Church and Christian Heritage College, galvanized fundamentalist fears of humanism by rendering a popular version of the Whitehead and Conlan thesis. LaHaye built on the Whitehead and Conlan definition of humanism that emphasized autonomy: "Simply stated, humanism is man's attempt to solve his problems independently of God," and "humanists view man as an autonomous, self-centered godlike person" (LaHaye 1980:26,68).

LaHaye illustrated the evils of humanism by referring frequently to pornography, homosexuality, drug addiction, abortion, and giving away the Panama Canal to Communists. The cumulative product was a low-brow Manichaeanism. LaHaye attested that "most of the evils in the world today can be traced to humanism" (9), and that "crime and violence in our streets, promiscuity, divorce, shattered dreams, and broken hearts can be laid right at the door of secular humanism" (26). (In LaHaye's book, "humanism" is shorthand for Secular Humanism, as if all humanism is equivalent to Secular Humanism.) Finally, LaHaye elevated two documents, the Humanist Manifestos I and II, to the status of the sacred scripture of the humanists: "What the

Bible is to Christians, the Humanist Manifesto is to humanists" (85).

The impact of LaHaye's book was startling. "In 1979 many fundamentalists had not even heard of secular humanism," wrote Jeffrey K. Hadden and Charles E. Swann. "It was not mentioned in sermons and writings. But by the end of 1980 nearly all had adopted it as their enemy" (Hadden and Swann 1981:86). Rev. Lamarr Mooneyham, the Moral Majority leader of North Carolina, said, "I wasn't aware of the growing influence of secular humanism before the early 1980s. But as I began to address issues facing Moral Majority chapters, at the root of every opposition was an active or passive humanist" (*News & Observer* [Raleigh, N.C.], 12 May 1985). According to *Newsweek*, 350,000 copies of *The Battle for the Mind* had been sold as of July 1981 (6 July 1981:48).

The enemies of Secular Humanism scored a victory when Senator Orrin G. Hatch, Republican of Utah, had a provision passed in 1984 to deny funding for the teaching of Secular Humanism. But the Hatch Amendment lacked a definition of the thing it decried, and it expired before having any apparent effect on public school education.

The legal strategy of Whitehead and Conlan dovetailed with LaHaye's sweeping moral critique in the case of *Smith v. Mobile*. Alabama had passed a law in 1982 encouraging school prayer directed by teachers, whereupon an agnostic parent named Ishmael Jaffree challenged the law, and the U.S. Supreme Court ruled it unconstitutional. A federal district court judge, W. Brevard Hand, a conservative appointed by President Nixon, then continued the case so that Christians might do to humanism what Jaffree had done to school prayer. Judge Hand engineered a class action suit in which "all those persons adhering by belief and practice to a theistic religion" could become plaintiffs for the purpose of alleging that Secular Humanism was a religion, and that it was being taught in Mobile public schools, contrary to the *Abington* stipulation (Hand 1987). Thus the trial became a platform for popular hostility to Secular Humanism. It consumed three weeks in October 1986. In March 1987 W. B. Hand delivered his judgment that, indeed, "for purposes of the first amendment, secular humanism is a religious belief system"

(Hand 1987), and that the contents of certain public school textbooks represented an establishment of that religion.

To construct a definition of Secular Humanism, Judge Hand heard testimony from numerous scholars. Many identified John Dewey as the most influential humanist thinker in the United States, whereupon the philosopher Russell Kirk explained that Dewey's brand of thought was known as *religious* humanism, and that *Secular* Humanism arose as a reaction against Dewey's grandiose vision. In other words, Secular Humanism constituted the view that humanism ought not to have the status of a religion, as the adjective plainly indicates. Nevertheless, Judge Hand wrote in his legal definition of Secular Humanism that "the most important belief of this religion is its denial of the transcendent and/or supernatural: there is no God, no creation, no divinity" (Hand 1987). Although he attributed some substantive content to Secular Humanism, he departed from the definition by Whitehead, Conlan, and LaHaye (which had identified Secular Humanism in terms of autonomy) and instead depicted this belief system principally as a negation of the supernatural.

In August 1987, a three-judge panel of the Eleventh U.S. Circuit Court of Appeals overturned Judge Hand, but the panel addressed only the narrow issue of his order regarding textbooks and avoided the larger question, namely, whether Secular Humanism is a religion.

What Is Secular Humanism?

In general, Christians have adopted two styles for expressing the meaning of Secular Humanism. I call these the "negation of personal beliefs" and the "autonomy theory." Consider these four examples of Secular Humanism as a negation of one's personal beliefs:

- The *Bible-Science Newsletter* lists nine components of humanism: naturalism, evolution, faith in humanity, faith in reason and science, relativism, situational ethics, anti-authoritarianism, civil liberties, and globalism (May 1984:7 insert).

84

- A tract titled "Humanism: America's Greatest Enemy" gives these features of humanism: "OK to lie, OK to kill, OK to steal, OK to have pre-marital sex, OK to cheat," plus transcendental meditation, yoga, witchcraft, masturbation, children playing the roles of homosexuals and unwed parents, survival games, communism, atheism, evolution, and amorality (Bolles n.d.).
- A letter to the *Charlotte* (N.C.) *Observer* states that "abortion, pornography, evolution, sex and values education, socialism, communism, and bureaucratic government are all part of secular humanism" (13 June 1980).
- A pamphlet titled "Is Humanism Molesting Your Child?" says that humanism includes the denial of these beliefs: deity of God, inspiration of the Bible, divinity of Jesus Christ, existence of the soul, life after death, biblical account of creation, and absolute standards of right and wrong; furthermore, it alleges that humanism embraces sexual freedom "regardless of age," plus incest, removal of male-female distinctions, control of the environment, "removal of American patriotism," disarmament, and, finally, "one-world socialist government" (ProFamily Forum 1980).

This last version has been quoted widely, often verbatim, though usually without attribution. In North Carolina, I noticed it in a local Moral Majority critique of textbooks from 1981 (North Carolina Moral Majority, 1981), in a letter to the editor in a small newspaper during a curriculum controversy (*News of Orange County*, 19 September 1984), and in the complaints of the state's severest textbook critic (*News & Observer*, 12 May 1985), to name but three instances.

Notice how this style of expression attributes substantive content to the term *Secular Humanism*. Its authors list their most precious spiritual values and their most troubling fears, then make Secular Humanism their mirror image, value for value and fear for fear.

A comparison of the four examples reveals another important fact. There is some overlap of particulars among the four, but many items are unique to one example or another. The first includes civil liberties but not masturbation; the second

85

mentions masturbation but not bureaucratic government; the third has bureaucratic government but not incest; and so on. Each author customizes his or her own portrait of Secular Humanism because there are no apparent rules regarding what is to be included, and what excluded. Nor does it require any pattern of relations among the particulars, or between the particulars and the general term. The cumulative effect is multiple solipsism. Secular Humanism means something different to each enemy of Secular Humanism.

The second style of defining Secular Humanism holds that atheism constitutes a vacuum of ethical values, which is then filled by an attitude of extreme human autonomy: we must be our own supreme being since there is none higher than ourselves. Finally, autonomy is said to lead to anarchy because each individual will live in a world of moral relativism and situation ethics, not recognizing any common standard of morality. Thus Secular Humanism is thought to be a slippery slope from atheism to autonomy to anarchy.

Recall that LaHaye's book and the article by Whitehead and Conlan emphasized autonomy in their definitions. The following passage from a creationist newsletter echoes those views:

When man substitutes his own knowledge and wisdom for the Creator's and allows every man to believe he is a law unto himself, there is no need to recognize the conscientious "absolutes" imposed on him by government. But where does this thinking lead? Can we as a nation survive if we believe in the absolute authority of men and deny the law of God? Can we allow the standard to be the lowest common level to which man can sink? (*Creation-Science Report*, December 1978)

Hostility to human autonomy also arises in the comments of creationism's grass-roots activists in North Carolina, who asserted that Secular Humanism constitutes:

A worship of man alone; man as the ultimate center of appreciation; man is in control of his destiny; the ex-

86

altation of man; man as the Supreme Being; a king-size ego trip; elevation of man to the God level; the captain of his destiny, apart from any deity; man himself determines what is right and what is wrong; a man-oriented frame of reference; a man-centered religion; man is the measure of all things; the enthronement of man; man is the center of the universe; man is not responsible to a superior being. [All comments are verbatim.]

These grass-roots expressions indicate how completely the Whitehead-Conlan-LaHaye autonomy definition has been accepted. By contrast, none of these interviewees in North Carolina described Secular Humanism according to the negation of beliefs approach, thus suggesting that it is a literary device reserved for the Religious Right's tracts, pamphlets, and letters to newspapers.

An important relation between the two definitions is that the autonomy theory is distinctly more focused than the negation of personal beliefs approach, for it offers hypotheses about philosophical linkages and social causation that can be tested, or at least debated. But the idiosyncratic expressions of discontent from the negation of beliefs style lack such clarity. For example, the autonomy theory states a hypothetical relation between atheism and autonomy, such that atheism is the ultimate cause of anarchy, whereas the proximate cause is autonomy. By contrast, the other approach carelessly lumps together atheism and autonomy, along with many other items. With this distinction in mind, we can weigh the impact of various statements about Secular Humanism. Consider again Judge Hand's opinion in the *Mobile* case, wherein he had defined Secular Humanism simply as a negation of theistic religion. The judge's findings of fact and conclusions of law add up to an unfocused anxiety in the negation of beliefs style. The same is true of Max Rafferty's definition in the "Guidelines for Moral Instruction" of 1969. As such, the views of Hand and Rafferty are less likely to become a useful legal definition of Secular Humanism than is the Whitehead-Conlan-LaHaye autonomy theory definition.

Either way, the term *Secular Humanism* gives a name to the deep moral outrage felt by fundamentalist Christians: when

87

The National Creationist Movement

you hear the phrase, you can expect it to be followed by a complaint about spiritual depravity in this land.

Cultural Realities of Secular Humanism

The views of Conlan, Whitehead, and LaHaye constitute a conspiracy theory that presumes that there is a distinct ideology named Secular Humanism, and that it has agents (the Secular Humanists); also, that these agents control much of U.S. culture, especially U.S. Supreme Court decisions and public school curriculums. This theory is much more dire than the observation that the nation's culture is becoming more secular as the Protestant hegemony is gradually retired (which indeed is happening). In fact, the conspiracy theory interprets all instances of secularization as having been spawned by a single highly organized force.

Indeed there is a certain modest reality to Secular Humanism. There is an American Humanist Association (AHA), and its Humanist Manifesto I, from 1933, is an explicitly antitheistic statement (*New Humanist*, May–June 1933). Its next proclamation, the Humanist Manifesto II, from 1973, is peppered with religion-baiting hostility, and it also makes claims about human autonomy, just as LaHaye and his colleagues charge. "Ethics is autonomous and situational, needing no theological or ideological sanction," it alleges (*Humanist*, September–October 1973). The humanism described in Paul Kurtz's definitive essay (Kurtz 1985) is truly a combination of atheism and individual freedom. To Kurtz, the history of humanism is the history of hostility to religion, with religion defined in terms of ignorance and superstition; that history culminates in the modern form called Secular Humanism. "Basically," says Kurtz, "secular humanists are atheists, agnostics, or skeptics" (330). Regarding autonomy, Kurtz writes that "the salient virtue [of humanism] is autonomy" (332). With those words, plus the two Humanist Manifestos, the AHA delivers to LaHaye more than enough material to build the argument that Secular Humanism can be reduced to atheism and autonomy.

As to humanism's status as a religion, one reference work states that "humanism is a religion with a nontheistic concept

88

of a supreme Creator" (Shulman 1981:398), and another says that AHA's certified humanist counselors "enjoy the legal status of ordained pastors, priests, and rabbis" (*Encyclopedia of Associations 1988*, 22d ed., s.v. "American Humanist Association").

These statements deserve to be put in perspective. The seven national humanist organizations have a gross total of 12,092 members (ibid.), although the net total is smaller if some groups have overlapping memberships. Compare these numbers with LaHaye's figure of 275,000 humanists (LaHaye 1980:179), which he established by adding a gratuitous 10 percent to an off-the-cuff estimate from the *New York Times* (26 August 1973), thereby giving the rough estimate the appearance of a precise quantity. In reality, the two Humanist Manifestos are obscure documents, better known to fundamentalists than to their enemies. There are barely enough Secular Humanists to populate a modest fringe movement, let alone control the moral climate of the nation's culture.

The humanists of the AHA hardly ever used the term *Secular Humanism* until after LaHaye popularized it. If not for an undefined term in a footnote to the *Torcaso* decision, this entity could just as well have been called Institutionalized Atheism, or Atheistic Individualism, or Programmatic Immorality, or any of a number of loosely interchangeable names. But because it acquired the name Secular Humanism, and because that term has been used interchangeably with the generic term *humanism*, it is common for fundamentalists to believe that humanism and secularism can be folded into one another.

And yet a rich tradition of Christian humanism coexists with the secular form; also, there are forms of secularism that are not at all humanistic. Quattrocento Italian Renaissance humanism was highly compatible with Christian belief (*Encyclopedia of Philosophy*, s.v. "humanism"; Guistiniani 1985:192–193). In fact, one of its offspring was Christian humanism, best represented by the writings of Erasmus. Its scriptural referents are the life of Jesus and the Eighth Psalm: "What is man, that Thou art mindful of him? And the son of man, that Thou visitest him? For Thou has made him a little lower than the angels, and has crowned him with glory and honour."

89

Spanning five centuries, the tradition of Christian humanism embraces the ministries of Martin Luther and Martin Luther King, as well as those of Albert Schweitzer, Pope John XXIII, Jacques Maritain, and Mother Teresa (*Encyclopedic Dictionary of Religion*, s.v. "humanism, Christian"). Even Sidney Hook, the arch-antitheologist, saluted Maritain's Christian humanism for opposing totalitarianism (Hook 1987:336–347).

The American Humanist Association is just as mistaken as LaHaye: while humanism is not intrinsically religious, neither is it intrinsically secular, let alone atheistic (Giustiniani 1985:192–194). And if the miniscule numbers of the American Humanist Association are seen in proportion, then it is impossible to accept either the claim that Secular Humanism governs all that is bad in Western civilization, or the counterclaim that it produces all that is good.

Likewise, secularism cannot be reduced to humanism. Stalinism and Nazi fascism were secular enemies of Judeo-Christian religion, but no humanist would claim that either was humanistic. The New Age movement is a secular alternative to Judeo-Christian spirituality, but it is deterministic in its astrology and antirationalist in its epistemology. Paul Kurtz has sensibly attempted to put as much philosophical distance as possible between his brand of humanism and the New Agers (Kurtz 1989).

And so, if some humanism is secular, but much else is not, and if some secularism is humanistic, while much else is not, then the fundamentalist conspiracy theory of Secular Humanism is a profound distortion of humanism, grounded in a great exaggeration of the modest reality of the American Humanist Association.

Irresponsible Molecules Indicted: Evolution Meets Secular Humanism

Because fundamentalist Christians use Secular Humanism as the framework for their understanding of immorality, they must incorporate the idea of evolution into Secular Humanism if evolution is to be tied to immorality. This is done in both styles of characterizing Secular Humanism. In the negation of personal beliefs approach, belief in evolution is often listed as one of the many items that collectively constitute Secular Humanism. The first three examples of that approach cited earlier included evolution. But the approach includes no clear principles for linking evolution to the general term. And even though most examples of the negation of beliefs definition include evolution, some others omit it, while emphasizing sexual issues or belief in communism or New Age thought and so on in their own inventories of Secular Humanism.

A more serious matter is evolution's purported relation to the autonomy theory. As Secular Humanism is said to be a process by which autonomy generates anarchy, so evolution is accused of promoting this process by implying that randomness in nature justifies anarchy in society. If this is so, then people who believe in evolution will think anarchy is good and natural. Thus, evolution is charged with being the natural history of anarchy. Note that the task of fitting evolution into that theory requires one to represent it as a celebration of randomness: evolution is described in terms of its stochastic features, especially mutation. In 1972 creationists in California accused evolutionists of teaching that "the universe, life, and men are simply 'accidents' that occurred by fortuitous chance without cause, purpose, or reason" (*New York Times*, 17 December 1972). Also in that year, Henry Morris wrote that "the very essence of evolution, in fact, is random mutation, not scientific progress" (Morris 1972:271). Later, an ICR writer stated that "Christ offers purpose and hope for eternity; evolution proffers randomness and uncertainty forever" (Morris and Gish 1976:315). In his 1978 lawsuit against the Smithsonian, Dale Crowley described evolution as "the assumption that man and all life on earth is the consequence of a series of accidents of molecular combination in the dateless past" (*New York Times*, 14 December 1978). Morris, in his history of creationism, alleges that evolution "necessarily means endless ages of random changes which, in the process, leave untold waste and pain and death in their wake" (Morris 1984:328).

So the problem with evolutionary thought is that if people believe molecules behave irresponsibly in nature, then they will believe it is all right for individuals to behave irresponsibly in society. The most succinct denunciation of this natural history of anarchy comes from R. L. Wysong: "If life came into existence through purely natural, materialistic, chance processes, then, as a consequence, we must conclude life is without moral direction and intelligent purpose. . . . Atoms have no morals, thus, if they are our progenitors, man is amoral" (Wysong 1976:6).

The same complaint resonates in the comments of some of the creationists I interviewed in North Carolina:

> If one accepts the evolutionist point of view . . . he's just a random product of molecular collisions. So man is answerable to himself, which I think is dangerous. He then has the freedom to set his own moral standards.

> The evolution model says life arose from nothing, by pure random chance. . . . This theory of origins is leading to the view that you're not responsible to a higher authority, that man is only responsible to himself. . . . Life is a totally random event. Nothing really happens.

> Secular humanism is the idea that man himself determines what is right and what is wrong, that he doesn't have to answer to any higher being. . . . An evolutionist who believes that things came about by chance would also say, we are not answerable to any higher being.

From these statements a simple parallel can be constructed: evolution is to Secular Humanism as random molecules are to irresponsible individuals. This view eliminates classic Darwinian thought from evolution. It draws attention to mutation and other stochastic processes of evolution, but it deletes evolution's deterministic features, such as adaptation and differential reproductive success, which together constitute the process of natural selection. In fact, conventional evolutionary thought considers evolution to be an interaction between stochastic and deterministic processes, as Ernst Mayr suggests:

As for the objection to the presumed random aspect of natural selection, it is not hard to deal with. The process is not at all a matter of pure chance. Although variations arise through random processes, those variations are sorted by the second step in the process: selection by survival, which is very much an anti-chance factor. . . . Selectionist evolution, in other words, is neither a chance phenomenon nor a deterministic phenomenon, but a two-step tandem process combining the advantages of both. (Mayr 1978:53)

Some opponents of creationism have objected to editing out natural selection. Norman Newell complained that, contrary to creationist representations, "most biologists now recognize natural selection as the directive force in evolution. No modern evolutionist believes that evolution is the result of a long series of random accidents" (Newell 1974). William Pollitzer, reflecting on his 1974 debate with Henry Morris, recalls that Morris "suggested that evolution must be equated with chance. Yet I see nothing in evolution that denies the laws of cause and effect operating in an orderly universe. . . . It is natural selection in its interplay with the changing environment that ensures direction, in contrast to the disorder implied by the word 'chance'" (Pollitzer 1980:329–330).

Why has this misrepresentation become fixed in creationist thought? I believe the answer is that creationist thought stems from fundamentalism's critique of immorality in U.S. society, as organized according to the concept of Secular Humanism. It must conform to the ideological structure of that critique, in particular to uphold the autonomy theory's accusations about anarchy, which means that creationism must amplify randomness and mute Darwinian determinism in its representations of evolution.

One final means of connecting evolution to immorality alleges that evolution is directly responsible for immorality, without reference to Secular Humanism. Earlier in this century, that was the way William Jennings Bryan denounced evolution, for example, when in 1922 he charged that evolutionists "weaken faith in God, discourage prayer, raise doubt as to a future life, reduce Christ to the stature of a man, and make the Bible a 'scrap of paper'" (Bryan 1922).

In the contemporary version of this accusation, Judge Braswell Dean of the Georgia Court of Appeals says: "This monkey mythology of Darwin is the cause of permissiveness, promiscuity, pills, prophylactics, perversions, pregnancies, abortions, pornography, pollution, poisoning, and proliferation of crimes of all types" (*Time*, 16 March 1981:82).

Likewise, Nell Segraves of the Creation-Science Research Center concludes that "the research conducted by CSRC has demonstrated that the results of evolutionary interpretations of science data result in a widespread breakdown of law and order. This cause and effect relationship stems from the moral decay of mental health and loss of a sense of well being on the part of those involved with this belief system, i.e., divorce, abortion, and rampant veneral disease" (Segraves 1977:17).

The most prolific and most vehement source of this view is Henry Morris of ICR, who asserts that "the deception of evolution" was responsible for Satan's rebellion against God, Eve's deception of Adam, and Satan's deception of the world (Morris 1963:93). Also, Morris suggests that Satan invented evolution at the Tower of Babel (Morris 1974:74–75). He writes that "the foundation of false teaching in every discipline of study, and therefore of ungodly practice in all areas of life, was evolutionism" (Morris 1984:223), and he presents specific examples:

> If man is an evolved animal, then the morals of the barnyard and the jungle are more "natural," and therefore more "healthy," than the artificially-imposed restrictions of pre-marital chastity and marital fidelity. Instead of monogamy, why not promiscuity and polygamy? . . . Self-preservation is the first law of nature; only the fittest will survive. . . . Eat, drink, and be merry, for life is short and that's the end. So says evolution! (Morris and Gish 1976:172)

Morris traces pagan religions, humanism, and the New Age movement to evolution (Morris 1982; 1983; 1987) and also holds it responsible for "most of the spiritual and moral problems that have arisen to hinder the gospel" (Morris 1984:352). Elsewhere, McIver (1989:294–302) and Harrison (1990) give additional examples of the accusation that evolution is directly responsible for immorality.

Blistering as are these diatribes, they fail to specify exactly how evolution could have caused so much immorality. In this regard they resemble the negation of personal beliefs definition. The "Evolution Tree" of Richard G. Elmendorf (see figure) presents a vivid image of that pattern. Elmendorf, an engineer whose adamant commitment to creationism is tempered with charm, patience, and good humor, represents evolution as the trunk from which twenty-one evil fruits grow, including dirty books, inflation, and terrorism. His drawing suggests that those things can be defeated by using creationism to destroy evolution.

Although it is impossible to quantify the authority of each of the three approaches (negation of beliefs, autonomy theory, and theory of direct responsibility), the autonomy theory is probably the most influential condemnation of evolution, for it offers a relatively clear plan by which evolution is said to be tied to the grand scheme of immorality named Secular Humanism. The other two approaches draw much attention by virtue of their lurid and sweeping allegations, but they fail to answer the central ideological question that excites modern creationism, namely, How is evolution connected to immorality?

A Creationist Sociology of Evolutionists and Humanists

If evolution generates immorality, directly or indirectly, then those who defend evolution are agents of immorality, intentionally or not. That idea establishes a series of denunciations of the moral character of evolutionists. Most frequent is the charge that evolutionists are "pompous and arrogant, just the kind of people that the First Amendment was written to protect us against," and that they display "an academic arrogance frequently typical of the nation's scientific-educational establishment" (*Science* 20 March 1981:1331–1332, 1 June 1979:925).

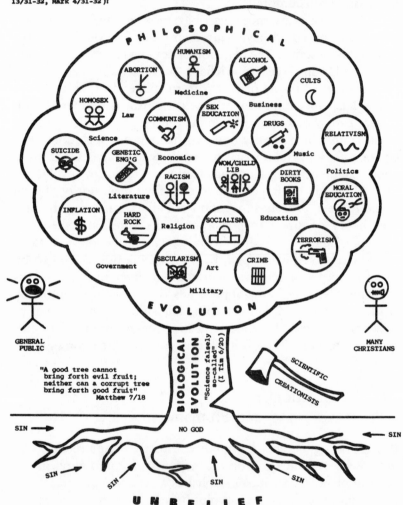

THE EVOLUTION TREE

Evolution is not just a "harmless biological theory" - it produces evil fruit. Evolution is the supposedly "scientific" rationale for all kinds of unbiblical ideas which permeate society today and concern many people. Evolution "holds up" these ideas, and if evolution is destroyed, these ideas will fall.

The "kingdom" of evolution starts with a small no-god biological idea sown in the fertile ground of unbelief. Nourished by sin, it grows up into a "tree", sending out "great branches" into education, government, medicine, religion and so forth, and eventually covering the "whole world". Then all sorts of evil ideas come and "lodge in the branches thereof" (Matthew 13/31-32, Mark 4/31-32):

PHILOSOPHICAL

HUMANISM · ABORTION · ALCOHOL · CULTS · HOMOSEX · SEX EDUCATION · COMMUNISM · DRUGS · RELATIVISM · SUICIDE · GENETIC ENG'G · WOM/CHILD LIB · DIRTY BOOKS · RACISM · MORAL EDUCATION · INFLATION · HARD ROCK · SOCIALISM · TERRORISM · SECULARISM · CRIME

Medicine · Business · Law · Science · Economics · Music · Literature · Politics · Religion · Education · Government · Art · Military

EVOLUTION

GENERAL PUBLIC

MANY CHRISTIANS

BIOLOGICAL EVOLUTION

"Science falsely so-called" (I Tim 6/20)

SCIENTIFIC CREATIONISTS

"A good tree cannot bring forth evil fruit; neither can a corrupt tree bring forth good fruit" Matthew 7/18

SIN → · NO GOD · ← SIN · SIN → · SIN · SIN · SIN

UNBELIEF

What is the best way to counteract the evil fruit of evolution? Opposing these things one-by-one is good, but it does not deal with the underlying cause. The tree will produce fruit faster than it can be spotted and removed. A more effective approach is to chop the tree off at its base by scientifically discrediting evolution. When the tree falls, the fruit will go down with it, and unbelieving man will be left "without excuse" (Romans 1/21). That is the real reason why scientific creationism represents such a serious threat to the evolutionary establishment!

Pittsburgh Creation Society
Bairdford Pa 15006
R G Elmendorf

The Evolution Tree (courtesy of R. G. Elmendorf)

Evolution and Secular Humanism

Evolutionists also are said to be categorically intolerant. The *Bible-Science Newsletter* of March 1975 tells its readers that "fools despise wisdom and instruction. This is true of evolutionists who refuse to consider evidence which disagrees with their preconceived ideas of age." The Institute for Creation Research proposes that biology teachers can be expected to grant scientific credibility to creationism "unless absolutely blind, or dulled by prejudice" (Morris, Gish, and Hillestad 1974:27). Elsewhere ICR cites a survey of network television executives, whom it reports as very tolerant of adultery, homosexuality, abortion, and liberalism; it concludes that, "although the creation/evolution question was not specifically addressed in the survey, it is obvious that such a group of people would be heavily biased in favor of evolution" (*Acts & Facts*, July 1983).

In addition, evolutionists are thought to be systematically deceitful. It is a regular feature of creationism to allege that the Java Man and Peking Man fossils are intentional frauds. The August 1985 *Creation-Science Prayer News* warns that "evolution traps" have "been set by atheistic humanists all over our country. . . . The traps are hidden in vacation and amusement locations to put the victims off their normal guard . . . [so that] the tourist . . . [inhales] large doses of evolutionary indoctrination, or even worse, he has parroted it to his children." Among the "evolution traps" it identifies are the Museum of Science and Industry and the Field Museum, both in Chicago, plus the Grand Canyon, as well as zoos and aquariums in general.

All the usual impeachments of evolutionists' moral character are captured concisely by Duane Gish of ICR in his closing remarks at the October 1981 debate at Liberty Baptist College. In two short paragraphs of text, Gish accused evolutionists of being dogmatic, intolerant, deceitful, arrogant, elitist, afraid of creationism's truths, afraid of majority sentiment, and accustomed to indoctrinating their students ("Old-Time Gospel Hour" 1981).

These judgments are especially vivid in the fundamentalist comic-book tracts of Jack T. Chick, a California publisher. In *Primal Man?*, a Christian anthropologist confronts a crew making an "evolution film." The director, named Dexter, is shown to be bearded, vain, and effeminate, dressed in a purple jumpsuit with a saffron scarf. His producer acknowledges that

97

evolution is wrong but decides to continue making his films, "even though it's brainwashing these kids. . . . Many will lose their souls because of these films" because, he admits, he worships money as his god (Chick 1976).

Another of Chick's comic-book tracts, *Big Daddy?*, depicts a professor of evolution as fat, bearded, and balding; he becomes hysterical when a student mentions the Bible. The other students, all of whom are evolutionists, include these figures: a black man with an Afro, dark glasses, and beads; a hippie woman with a fringed vest and long hair; a man with a peace medallion; and one angry long-haired man making a clenched-fist sign and another making a V-sign for peace. Thus, into the first two panels, the artist has packed six right-wing caricatures of leftists, associating them with evolution (Chick 1972).

Rev. Tim LaHaye's book, *The Battle for the Mind*, tells the reader that the humanists' allies are "the moral degenerates, such as gangsters, prostitutes, porno publishers, dope peddlers, some homosexuals, and others without any moral values" (LaHaye 1980:184), and he uses the term "amoral humanists" to identify "the legislators, judges, governors, attorneys-general, and so forth who have created the laws that enable porno publishers to pollute the minds of our young" (173).

Occasionally real people are denounced by name (most often Isaac Asimov, Stephen Jay Gould, Carl Sagan, Corliss Lamont, and Paul Kurtz), but the more common pattern is to deride evolutionists and humanists in terms so sweeping and so general that all of one's enemies are interchangeable. Evolutionists are categorically arrogant, or all humanists are equally depraved. The creationist commentary on evolution and humanism is a Manichaean ideology in which two sets of moral abstractions struggle against each other to control the nation's culture: autonomy versus piety, immorality versus biblical belief, arrogance versus humility, and deceit versus perspicuity. In other words, spiritual character and abstract virtues are thought to determine the collective destinies of humanism, evolution, creationism, and fundamentalist Christianity.

Creationism, then, is much more than a narrow doctrine extrapolated from a handful of biblical verses. It represents a broad cultural discontent, featuring fear of anarchy, revulsion

98

for abortion, disdain for promiscuity, and endless other issues, with evolution integrated into those fears. Whether one agrees or disagrees with creationist moral theory, it should be recognized that this is a rich and complicated understanding of U.S. culture that gives considerable existential depth to conservative Christian understandings of reality.

Six
The Core of Scientific Creationism: Organizations and Ideologies

As a body of thought, scientific creationism combines an understanding of science with a moral theory. Its stance on science, broadly stated, is that the plenary authority of science rightfully belongs to the Protestant model of science, with the implication that science corroborates the literal, historical authenticity of the narratives in Genesis. But just as creationism involves multiple moral interpretions of evolution, so there are also multiple creationist interpretations of scientific authority. Those latter interpretations emerged from the institutional histories of the principal creationist organizations. To understand the various creationist positions on science and scientific authority, it is necessary to consider the histories of those groups.

A series of organizational decisions in Southern California between 1969 and 1972 established the basic relations among the three core creationist groups, which are the Creation Research Society, the Institute for Creation Research, and the Creation-Science Research Center. In 1963, the only genuine creationist organization was the Geoscience Research Institute (GRI) of Andrews University in Berrien Springs, Michigan, which served only the Seventh-day Adventist Church. At that time, most creationists in the scientific community belonged to the American Scientific Affiliation.

ASA's tolerance of evolutionary thought distressed its hardline creationist members. Walter E. Lammerts, a plant geneticist who was "disgusted by ASA's flirtation with evolution," began to correspond with like-minded ASA members during the early 1950s (Numbers 1982:542; Rusch 1982). When Whitcomb and

100

The Core of Scientific Creationism

Morris's book, *The Genesis Flood*, appeared in 1961, it pleased creationists to see the argument made that science supports biblical creationism. In an effort to institutionalize and increase that claim on scientific authority, a group of Morris's colleagues and admirers met in a creationist caucus at the June 1963 meeting of ASA and there established the Creation Research Society. The founders quickly hammered out a confessional statement that separated this organization unambiguously from ASA, lest the new group perish in what Henry Morris called "a quicksand of pseudo-creationist evolutionism" (Morris 1984:181–184, 197; Rusch 1982). The CRS statement, which voting members, sustaining members, and student members must sign, presents these four beliefs:

1. The Bible is the written Word of God, and because we believe it to be inspired thruout [*sic*], all of its assertions are historically and scientifically true in all of the original autographs. To the student of nature, this means that the account of origins in Genesis is a factual presentation of simple historical truths.

2. All basic types of living things, including man, were made by direct creative acts of God during Creation Week as described in Genesis. Whatever biological changes have occurred since Creation have accomplished only changes within the originally created kinds.

3. The great Flood described in Genesis, commonly referred to as the Noachian Deluge, was an historical event, worldwide in its extent and effect.

4. Finally, we are an organization of Christian men of science, who accept Jesus Christ as our Lord and Savior. The account of the special creation of Adam and Eve as one man and one woman, and their subsequent Fall into sin, is the basis for our belief in the necessity of a Savior for all mankind. Therefore, salvation can come only thru accepting Jesus Christ as our Savior. (Morris 1984:339)

The first of these was a proclamation of biblical inerrancy, and the fourth was an evangelical declaration that humankind cannot be saved from sin except by accepting Jesus. The second was a traditional creationist statement of polygenism, meaning that biological diversity represents numerous independent acts of creation by God, as opposed to speciation via evolutionary processes. The third point showed the special influence of Henry Morris: creationists could collapse all the power of catastrophism into one special catastrophe, Noah's Flood. In traditional catastrophism under Georges L.C.F.D. Cuvier and J.L.R. Agassiz, this flood was only one in a series of great catastrophes. That way, creationists could accept long geological periods with numerous changes connecting them. In Morris's modification of catastrophism, however, Noah's Flood accounted for all major features of the geological record. Inferring a great age for earth history thus became heretical. So strict was the Creation Research Society that its quarterly journal pledged never to publish articles that advocate "old earth" beliefs (Morris 1984:192).

Having declared its doctrinal position, CRS announced that its primary purpose was the publication of the *Creation Research Society Quarterly*. The journal ordinarily includes both technical and exegetical articles, but its specialty has been to combine the two genres, using technical arguments to support biblical inerrancy and biblical insights to understand technical problems. James E. Strickling's article, "A Statistical Analysis of Flood Legends," illustrates this combination of styles. The author surveys forty-seven flood legends from around the world and classifies them according to presence or absence of such elements as "a favored family is saved," "a remnant survives," "survival is due to a boat," and so on. Then, using chi-square tests to determine which elements correlate most frequently, he infers that the strongest cluster of elements represents an original flood story from which others are derived, and that weaker clusters of elements are incomplete derivations from the archetype. "A statistical analysis," concludes Strickling, "indicates the purity of the Biblical account and reveals evidence of subsequent upheavals having corrupted in varying degrees all other accounts" (Strickling 1972). Thus, science is said to

The Core of Scientific Creationism

substantiate the Bible, and the Bible is used to explain the history of flood legends.

In 1970, CRS published a high school biology textbook titled *Biology: A Search for Order in Complexity* (Moore and Slusher 1970). Four years later, the Georgia State Board of Education approved a revised edition of this book for public school use. (The Atlanta School Board rejected it with the comment that it "contains numerous errors in terms of established biological fact," and a superior court judge in Marion County, Indiana, considered its use an unconstitutional establishment of religion [*New York Times*, 23 June 1974, 18 April 1977]).

About a third of CRS members have advanced degrees in scientific fields, which leaves the society vulnerable to accusations of being unscientific or of being steered by nonscientists. To protect itself from those criticisms, it reserves voting rights for those with advanced degrees. In addition, the Creation Research Society is unaffiliated with any particular religious organization. Furthermore, the society keeps the names of its members confidential to protect them from the enemies of creationism (Morris 1984:187; Rusch 1982). Between 1963 and 1983, CRS membership grew from an initial 10 members to a peak of 701 members in 1982, then declined to 597 in 1983 (Morris 1984:194; Toumey 1987:84).

CRS considers itself strictly nonpolitical. In a brief description of its activities, it describes itself as "solely a research and publications society. It does not hold meetings or engage in other promotional activities, and has no affiliation with any other scientific or religious organizations" (CRS 1972; Morris 1984:187; Rusch 1982). Several CRS members participated prominently in the 1969 California science framework dispute (Grabiner and Miller 1974; Morris, Gish, and Hillestad 1974:13; Newell 1974), thus giving CRS a misleading reputation as a lobbying organization. Those were the acts of individual members, however. The two men most responsible for the creationist content of the California Science Framework, John R. Ford, chair of the state Board of Education, and Vernon Grose, an aerospace engineer, apparently had no CRS connections.

In 1970, Nell Segraves, Henry Morris, and Rev. Tim LaHaye joined forces to release creationism from the staid confines of

CRS's political and denominational neutrality. Segraves was a veteran anti-evolution activist in Southern California who had developed a novel legal assault on evolution after the U.S. Supreme Court's 1962 and 1963 decisions on prayer and Bible devotionals in the public schools. Whereas most conservative Christians were demoralized by those judgments, Nell Segraves was inspired by them. I interviewed her on 5 December 1983 at the Creation-Science Research Center in San Diego. "I thought," she said, "that Madalyn Murray [an atheist activist] had achieved something that everyone should have. We should have equal rights and equal representation and equal protection of our religious beliefs in the public sector." If atheists could eliminate compulsory prayer and Bible devotionals from the public schools, then, thought Segraves, Christians should be able to eliminate atheistic practices. This required that atheism be defined as a religion, not just as an absence of religion. Segraves successfully petitioned President John Kennedy to get a legal opinion from the attorney general to the effect that it would be unconstitutional to teach atheism as a religion in the public schools. Armed with this ruling, said Mrs. Segraves, "We wanted to identify what was atheistic. . . . So we chose evolution as an example of atheism because it was so prevalent in all the textbooks at that time."

When the California State Board of Education decided to include creationism in the science framework in 1969, Segraves and her son Kelly recognized an immediate need to publish creationist textbooks. This, in her view, was the most important reason for establishing the Creation-Science Research Center at Christian Heritage College, under the leadership of Henry Morris, in 1970.

Morris, meanwhile, had observed "an amazing revival of solid Biblical creation doctrine," so amazing that "we may be on the threshold of a sweeping movement back to faith in the integrity of God's Word." It was time, he felt, to establish "a nerve-center" of creationist sentiment (Morris 1984:352). Morris's idea coincided with LaHaye's project of starting a Christian college in the San Diego area.

Segraves, Morris, and LaHaye established the Creation-Science Research Center as the research division of Christian

The Core of Scientific Creationism

Heritage College in 1970. Morris became the director of CSRC and the college's vice-president for academic affairs. Kelly Segraves became the assistant director of CSRC, and Nell Segraves was its librarian. LaHaye was president of the college (Morris and Rohrer 1981:6). This way, the Segraves family had a publishing organization, Morris had a research institute, and LaHaye had a college organized around the concept of creation. Morris drew up the college's fourteen-point faculty doctrinal statement, in which its fundamentalist moral theory was centered on creationism, so that "all curricula [are] to be founded on creationism and full Biblical authority" (ibid.:222).

The Christian Heritage College statement was a remarkable document. The brief generic suppositions of the Creation Research Society's 1963 pledge paled next to Henry Morris's 1970 masterpiece, which was both comprehensive and precise in its outline of individuals' relationships to each other, to God, and to evil. In four somber, detailed pages, Morris mapped the central features of creationist belief and reviled the enemy's assumptions. According to the 1963 CRS pledge, the critical scriptural proof-text was simply "the account of origins in Genesis"; in the 1970 CHC statement, it was specifically Gen. 1:1– 2:3, which exalted the six-day narrative and disregarded the account in which the Creator formed Adam from dust, and Eve from Adam's rib (Gen. 2:7, 2:21–22). The first creation account was "foundational to the understanding of every fact and phenomenon in the created universe." Evolutionary thought "in any form" was categorically rejected (Morris 1984:355–359). Four of the fourteen points delineated the individual's dependence on Jesus Christ for personal salvation through Christ's substitutionary atonement, followed by his bodily resurrection, and soon to be followed by his imminent return; all this displaced the brisk evangelical creed in the CRS pledge. Morris's CHC statement then recapitulated the existential assertions that evil is real, and that God permits it in his plan so that individuals can exercise free will by choosing between good and evil. Evil, however, was not a passive impersonal force in this doctrine. It was an unlimited conspiracy in which "all evil in the universe is headed up in the cosmic rebellion instigated by . . . Satan" (357). The reign of evil will cease at the end of time, it said, whereupon Satan and his followers will dwell eternally in a lake of fire. There will also be room in this lake for everyone who has declined to be saved by Jesus (358).

After explaining God, creation, sin, redemption, and evil, Morris outlined human history in terms of seven biblical events. Noah's Flood, for example, accounted for changes in the face of the earth, while the Tower of Babel marked the origin of multiple languages. Finally, the author extrapolated the individual's relation to society: family, church, and government are the basic units of God's plan for human society. This complicated document conceded absolutely no philosophical deviation and no interpretive variation.

Schism, however, erupted within CSRC. Within about a year of its founding, the Creation-Science Research Center experienced a painful internal dispute that eventually split much of the staff from the college and divided its creationist leaders into two inimical groups. This dispute was not in any sense doctrinal. Both parties still concurred on the central beliefs of creationist thought. They argued instead about tactics and strategies for disseminating creationist belief, with the Segraves group preferring expeditious political action, and the Morris group emphasizing gradual educational processes. The Segraves approach required denominational neutrality, but Morris wanted creationism to be situated within a church-affiliated college. This difference became a problem when Vernon Grose, representing the California State Board of Education, visited CSRC to examine the textbooks it had published in the summer of 1970. Grose observed that CSRC was a part of Christian Heritage College, which in turn was controlled by Scott Memorial Baptist Church, whereupon he had to reject the CSRC materials on the grounds that public school materials must come from nondenominational sources. Nell Segraves recalled this with regret in our 1983 interview: "Therefore we lost what might have been a real victory for California. So it was an uphill battle every year after that to overcome the obstacle of that association [with CHC]."

As they developed their political strategy, the Segraveses realized that if they were going to litigate for creationism and enjoy tax-exempt status, then their organization would have to

106

register as a public foundation, independent of religious identification (*Christianity Today*, 18 April:50, 7 November 1980:67). This decision obviously violated Morris's and LaHaye's vision for their creationist institute. Consequently, as Duane Gish explained to me in an interview at ICR in El Cajon, California, on 6 December 1983, "Problems began to arise between Mr. [Kelly] Segraves and Dr. Morris. Questions of procedure, and the Segraveses were more interested in the political aspects, promoting legislation and court tests and things like this, [while Morris and LaHaye] were more interested in the academic aspects."

In April 1972, the directors of CSRC voted eight to four to separate that group from Christian Heritage College. Morris, Tim LaHaye, Beverly LaHaye, and Art Peters (another officer of CHC) represented the minority position, to retain the relationship with the college. When the Segraveses left (and took the CSRC name with them), Morris and LaHaye decided at a prayer meeting to reestablish a creationist institute within CHC (Morris 1984:232–236). The new unit was the Institute for Creation Research.

After the schism, Henry Morris dedicated ICR to his own plan of educational activities uncluttered by political activism. In its first few years, however, ICR had to carve out an identity apart from CSRC's, which it did by criticizing CSRC directly but without naming it. For example, an early edition of *Acts & Facts*, the ICR monthly newsletter, cautioned that "many people have confused ICR with another group, but this impression should definitely be corrected" (Morris, Gish, and Hillstad 1974:72). "Another group" was obviously CSRC, but ICR preferred not to name it. In another *Acts & Facts* piece from about that time, Henry Morris delivered a forceful critique of CSRC's approach. He cautioned that even if creationists achieved favorable legislation or judicial decisions, these might not be implemented. Many teachers did not know how to teach scientific creationism; schools lacked creationist textbooks; the media might "unleash a barrage of journalist [*sic*] bombast and ridicule that could well destroy all the gains achieved by creation scientists in recent years." After recapitulating the principal issues dividing ICR from CSRC, Morris left CSRC and its followers

107

The National Creationist Movement

unnamed. This time they were citizens whose "politico-legal actions, however well-intentioned, are ill-advised and may do more harm than good" (ibid.:29–31).

The tension between ICR and CSRC sometimes generated unkind comments, as when ICR denigrated its counterpart as "a promotional and sales organization" and called Kelly Segraves's honorary degrees "false titles," or when Duane Gish, opposing hasty lobbying, complained that the lobbyists "were not that patient" (Nelkin 1982:82; *Christianity Today*, 22 January 1982:29). Largely, however, the two organizations concentrated more on their own respective responsibilities than on denouncing each other. When I interviewed Morris at ICR 6 December 1983, he said that, since their split in 1972,

> we've had really no connection with them [CSRC]. Occasionally we communicate, but even though we're in the same city, we really go our separate ways. They do their work, and we do ours. So we don't really know too much about what they're doing next. Our beliefs are essentially the same; maybe our methods are different. In terms of what we believe about the Bible, and about creation, and evolution, and so on, I think we're very close. They have tended to . . . concentrate more on political and legal types of involvement. And we have concentrated more on the educational and scientific aspects.

Thus the distinctions between CRS, CSRC, and ICR crystallized between 1969, when Nell Segraves and others recognized that CRS was unable to capitalize on the California educational decision, and 1972, when CSRC and ICR declared their different missions. We can now turn to more detailed considerations of CSRC's and ICR's roles in the movement.

The Creation-Science Research Center

CSRC pursued an aggressive legal attack on evolution after freeing itself from Morris's educational gradualism. Nell Segraves and her associates reasoned that the Christian child had

108

the same rights as the atheist child (Creation-Science Research Center 1975; N. Segraves 1977), so that "if it is unconstitutional . . . to teach of God in the public schools, it is equally unconstitutional to teach the absence of God" (K. Segraves 1984). Lest it seem that anything that fails to mention God is equivalent to teaching the absence of God, it should be realized that CSRC was pursuing both of the religion clauses in the First Amendment, with approximately equal emphasis. The Establishment Clause, forbidding government from favoring any religion, was the basis for the 1962 and 1963 Supreme Court decisions on group prayer and Bible reading in the public schools. The Free Exercise Clause prohibits government from interfering with religious behavior. Thus, government must not be in favor of any religion, but it must not be against any religion, either. While CSRC intended to use the Establishment Clause against atheists to forbid evolution, just as atheists had used it against conservative Protestants to forbid compulsory prayer, CSRC also felt that the Free Exercise Clause might be a lever for reestablishing Christian belief in the public schools. Kelly Segraves argued that, "since God's law and God's word are revealed in the Bible, programs which eliminate God or any reference to God and His word cause offense and a violation of the free exercise rights of Christian believers who hold these doctrines as truth" (K. Segraves 1984).

Kelly Segraves elaborated these charges by alleging that such a free exercise violation forces the Christian child to choose between the family's beliefs and the public school's teachings; also, that it favors the atheist child over the Christian child; and, most seriously, that it forces a Christian to sin by not putting God first (ibid.).

The CSRC activists no doubt realized the difference between their Establishment Clause arguments and their Free Exercise Clause contentions; the contradictions in CSRC's policies perhaps represented a transition from an earlier Establishment Clause strategy to a later free exercise defense of Christian belief. By the late 1980s, the free exercise argument had become more prominent in CSRC's program. Both approaches have had their frustrations for CSRC. The courts did not accept that evolution is a religion in the sense of the Establishment

109

Clause, yet neither did they agree with CSRC's free exercise reasoning.

Regardless, CSRC stayed busy with its lobbying. It participated in the unsuccessful defense of Tennessee's 1973 Equal Time for Genesis law; it sued to require that the Biological Science Curriculum Study return publishing revenues to the federal government; its Creation Creed Committee threatened to monitor local school boards in their presentation of evolution; and it dabbled in opposing sex education. *Christianity Today* reported that CSRC's "textbook vigilantes" had some success in chasing "dogmatic evolution" from science textbooks, and CSRC insisted that its textbook appraisals had generated "much more recognition given to the beneficial influence of Judeo-Christian beliefs on world cultures and civilization" (Creation-Science Research Center 1975; N. Segraves 1983; *Christianity Today*, 7 November 1980:64, 15 February 1982:70, 10 April 1981:60–61; *Creation-Science Report*, February 1978, June–July 1978).

In our 1983 interview, Nell Segraves described CSRC's methods candidly:

> We take a page from the Communists' book. You take two steps forward and one step back. We set out to influence other organizations, feeling that if you can influence one person at the top, they in turn can influence others. So our format is then to reach other organizations, and through them, their constituents. . . . We see ourselves as a guinea pig in the community. . . . If we can accomplish a particular goal, . . . then we'll turn around and recommend that to others.

The most elegant example of CSRC's strategy was the case of *Segraves v. California*, a lawsuit that came to court in March 1981. The Segraves family presented a story in which Kasey Segraves, Kelly's son, came home from school complaining that his teacher had told him he had evolved. Kelly sued the state Board of Education, "charging that his three children have been deprived of their religious rights because their belief in Creation as mankind's beginning has been superceded in science classes

The Core of Scientific Creationism

by the evolutionists' version." CSRC's attorney, Richard Turner, added that, because the state was hostile to religion, "we are seeking protection for the right to believe in a cause. The real issue is religious freedom under the First Amendment of the Constitution." Segraves and Turner were careful to state repeatedly that their case was *not* a matter of science versus religion but, rather, a freedom of religion question (*New York Times*, 3 March 1981; *Science*, 20 March 1981:1331; *Greensboro* [N.C.] *Daily News*, 5 March 1981).

The case attracted enormous media attention, not only from newspapers and newsmagazines, which predictably christened the case "Scopes II," but also from several science periodicals. The state Board of Education organized a phalanx of scientists to defend the educational merits of evolution. CSRC then withdrew almost all of its complaints, asking only that the judge reaffirm the 1973 Board of Education policy stipulating that evolution not be taught dogmatically. This left the judge slightly chagrined and the journalists sorely disappointed. When the court granted Segraves's innocuous request, there was nothing left to dispute. It is possible that CSRC had originally intended a major assault on evolution but retreated when the opposition's experts appeared. It is just as plausible, however, that the original complaint was a pretext to draw attention to the issue, and that CSRC had shrewdly expected only to consolidate the educational policy on dogmatism by evoking a judicial decision. Kelly Segraves commented afterwards that the judge's ruling "will stop the dogmatic teaching of evolution and protect the rights of the Christian child" (*Time*, 16 March 1981:80–82). Thus the final outcome of the case fit neatly into CSRC's freedom of religion strategy. (Whether Kasey Segraves's teacher, or any other teacher, was really teaching evolution dogmatically was never resolved.)

The Creation-Science Research Center committed itself so exclusively to its freedom of religion assault on evolution that it isolated itself from the primary goal of the rest of the creationist movement, which has been to have creationism taught as science in public school science courses. This difference left CSRC unable to participate as a principal in other major cases, such as the defense of the Arkansas law (*Christianity Today*, 4

September 1981:56), that are based on the equal-time approach. These cases required creationists to deemphasize religious issues. Denouncing evolution as atheism only detracted from the effort to present creationism on its scientific merits, for it drew attention to creationism's sectarian spirit. As she made clear in our 1983 interview, Nell Segraves was unimpressed by this reasoning:

> Our organization in particular never wanted to have a science-versus-science battle in the classroom. So, when we went to court on our lawsuit here [in California], we had a very difficult time keeping that particular argument out of our court case. We wanted to keep it on a constitutional basis only: the right of the Christian child and the defense of the Christian child against offense. If you get science versus science, there's no constitutional protection for that. Then you're arguing one man's opinion against another man's opinion, and most often the judge will go with the majority opinion. Not because it's right or wrong, [but] because it's all he knows. . . . Who wants to put a judge in that position? He can't decide between creationists and evolutionists. He's not a scientist. . . . I think [the Arkansas litigants] would have had a better position had they said that the manner in which evolution is presently taught is an offense to the Christian child and an evidence of bias.

The Institute for Creation Research

After CSRC left Christian Heritage College, Henry Morris and his colleagues reorganized as the Institute for Creation Research, again with an intimate relation between the creationist center and the college. But when ICR was planning its graduate school and applying for accreditation from the California education authorities, its intimate link with CHC became a liability, so the two institutions agreed to a formal separation. (Ironically, this decision lent support to Nell Segraves's original contention;

as she might have predicted, the close connection between ICR and CHS undermined ICR's scientific image.) ICR opened its own campus in Santee, a few miles from CHC, in January 1986. Formal separation seems not to have diminished CHC's creationist sentiment; *Acts & Facts* reminded its readers that CHC is still "unreservedly committed to strict creationism," and that "although the Institute for Creation Research became an independent educational institution in 1980, it has continued to maintain a close contractual working relationship with the College" (Morris 1984:229; Morris and Rohrer 1981:297; Gish 1983; *Acts & Facts*, March 1976, April 1984).

ICR is the most important messenger for bringing creationism to the public. Hundreds of thousands of Americans learn about the creationist case by attending debates, lectures, sermons, or seminars led by ICR's much-traveled speakers. These apostles of creation-science, bearing respectable, legitimate doctorates from major U.S. universities, carry their case to conservative churches, to conventions of Christian organizations, to public and private universities, to teachers' meetings, and to scientific conferences. Duane Gish, Henry Morris, and their ICR colleagues have changed the constitution of the evolution-creation quarrel from the old ways of the 1860 Oxford debate between Bishop Wilberforce and T. H. Huxley, or the Scopes trial of 1925 that set William Jennings Bryan against Clarence Darrow.

The creation-evolution debates mean much more to creationists than mere oratorical competitions; these are solemn symbolic confrontations that focus all considerations into exciting self-contained dramas, so that each debate, if it goes right, can be both its own little Armageddon and a rehearsal for the big Armagaddon to come. A debate presents a stark choice between two opposite positions. If one is right, the other must be wrong. If one side is victorious, the other has been humiliated. In their debates against evolutionists, ICR's speakers intend no compromise, no synthesis, no middle-ground resolution of these issues. Every debate is crucial. Just one loss for ICR could render its policies brittle and suspect. To contain this danger, ICR never admits to losing a debate. Its monthly newsletter articles on the debates always represent the creationists as convincing, reasonable, scientific, and well in command of objective facts. They "document" this, they "explain" that, they "easily counter" their opponents' assertions. In their most common idiom for describing themselves, they always "point out" the facts, "point out" the issues, and "point out" the problems. Evolutionist debaters, by contrast, can only "claim that" their arguments are true, according to the newsletter's corresponding idiom for its enemies, who "repeat old clichés," "speak in vague generalities," and "attempt to circumvent the difficulty" of creationist charges. Rather than support their claims, says *Acts & Facts*, the evolutionists' "disappointing presentations" "touch superficially" on the issues with "undocumented assertions" having "hardly a shred of scientific evidence." When they seem to do well, it reports, this is the result only of insidious tactics like "speaking with machine-gun rapidity." In short, ICR uses a formula that always ascribes credibility and success to its own speakers, as it dismisses the idea that evolutionists might ever win a reasonable debate.

One notable debate stretched this prose style to its seams. Kenneth Miller, a professor of biology at Brown University, debated Henry Morris in Providence, Rhode Island, on 10 April 1981. Disinterested listeners agree that Miller won the debate, at least by any ordinary standards. A woman who witnessed the debate, and who is an evangelical Christian, told me that Miller unquestionably won. Even *Acts & Facts* (June 1981) acknowledged that Miller was "an exceedingly knowledgeable and smooth speaker, the most effective evolutionist debater Dr. Morris has encountered to date." Five months later, when Morris met Miller in a second debate, ICR conceded that "Dr. Miller is a very personable and capable speaker, highly knowledgeable on the issues" (*Acts & Facts*, November 1981). Three years later, Henry Morris less generously called Kenneth Miller "an effective demagogue on the platform" to explain why Miller had done so well against him (Morris 1984:320).

For modern-day creationists, the painful memory of Clarence Darrow's triumphant cross-examination of William Jennings Bryan at the Scopes trial in 1925 has shadowed their own debates. Today's creationist debaters must overcome popular expectations that evolutionists will outclass them, and, more

114

importantly, that evolution is more intellectually appealing than the Holy Bible.

No single creation-evolution debate, regardless of how glorious or dismal, could undo what the Bryan-Darrow confrontation did. Nevertheless, ICR achieved a set-piece victory that more or less expiated the 1925 disaster and let the ghost of William Jennings Bryan rest in peace. Duane Gish, ICR's premier debater, met Russell Doolittle, a biochemist at the University of California, San Diego, in a debate at Rev. Jerry Falwell's Liberty Baptist College in Lynchburg, Virginia, on 13 October 1981. Both parties agreed that Falwell's television program, the "Old-Time Gospel Hour," would tape the debate and broadcast it later. The timing was ideal for creationists: the cause was in the news almost every day, attracting support subsequent to the Segraves trial in March, the Arkansas law of the same month, and the Louisiana law in July. Excitement was rising as the Arkansas law approached its December trial.

Doolittle and Gish had debated previously, though not for television. With Falwell as the moderator and his college as the audience, Gish had the advantage of talking on his own turf. Doolittle gave an interesting rambling talk that was fine as a college lecture but unorganized as a television presentation. Gish, the T. H. Huxley of creationism (Numbers 1982:543), was smooth, telegenic, and perfectly timed. ICR could justifiably claim victory, and, even better, it had the victory on videotape. ICR showed its pleasure and confidence by simply advising its audience: watch it on TV, and decide for yourself who won (*Acts & Facts*, December 1981).

ICR paces the movement with its seminars and lectures, both at Christian schools and at secular institutions. It sponsors summer schools and a radio ministry. The ICR staff has written more than sixty books on the issue of origins, and its publishing operation, Master Books (formerly Creation-Life), distributes filmstrips, tapes, and videotapes bearing the message. The institute has a modest Museum of Creation and Earth History, where visitors can see such material as a scale model of Noah's ark, modern artifacts encrusted in limestonelike calcium to show how quickly rocks can form, and miscellaneous artifacts of biblical archaeology. ICR boasts the only graduate school of creation-science (although CSRC hopes someday to have a "Genesis University" that "would attract the greatest minds in Christendom" [*Creation-Science Report*, March 1979]). ICR's research and publishing concentrates on combing evolutionary literature to expose problems, contradictions, and internal disagreements, but it also sponsors expeditions to Mount Ararat in Turkey to find Noah's ark, to the volcano Mount St. Helen in Washington State to observe the effects of a catastrophic geological event, and to the Grand Canyon to interpret its geological formations in terms of flood geology.

The Institute for Creation Research is particular about its unique position within the creationist movement, for it repeatedly reminds its readers that it must not be confused with, or even associated with, other groups.

ICR has no connection with the Creation Research Society, the Bible-Science Association, the Evolution Protest Movement, the Creation Science Association in Canada, or with any other creationist organization. We do appreciate the efforts of all sound creationist groups, of course, and co-operate with them whenever feasible, but our own program and organization are uniquely needed today and, we believe, have been raised up by God just for such a time as this. (Morris and Rohrer 1981:297–298)

The most important distinction is that ICR is a full-time permanent professional team. While it is not the oldest creationist group, it nevertheless represents "the first time a significant body of scientists and their support team have come together on a full-time basis to do research, writing and teaching in scientific Biblical creationism" (Morris and Rohrer 1981:6).

Above all, ICR emphasizes that it is *not* its crosstown cousin, the Creation-Science Research Center. Confusing the two is apparently chronic, at least for those who are not connoisseurs of creationism's internal subtleties. In 1980, Henry Morris had written that "ICR is recognized by all the news media (and especially by the opponents of creationism) as the real nerve

The Core of Scientific Creationism

center of the movement" (ibid.:7), but after CSRC generated enormous attention from the 1981 case of *Segraves v. California,* ICR had to remind its readers that the news media had erroneously linked ICR to the lawsuit. ICR expressed its disdain for the CSRC legal strategy without mentioning either CSRC or the Segraves family, and it concluded with a vague threat of libel action "if newspapers and periodicals continue printing untrue and defamatory statements about ICR and its staff" (*Acts & Facts* May 1981). The allegation of defamation did not explain how the media had gone from recognizing ICR so prominently to misidentifying it maliciously. (CSRC presumably did not share the opinion that confusing a group with CSRC was a defamation.)

ICR reaches its followers primarily through its monthly mailings, which include *Acts & Facts* (the newsletter that reports debates, lectures, and other events), *Impact* (a tract of four to eight pages on technical, educational, and doctrinal issues), and a cover letter that modestly solicits funds. ICR builds its mailing list by collecting names and addresses of interested people at debates, lectures, and seminars. The monthly mailings are free. The list is cleaned about every two years, so that ICR's constituency of seventy-five thousand or more regular readers constitutes a solid following seriously interested in its brand of creationism. No other group dedicated to creationism has a list even half that size.

The reader survey circulated by *Acts & Facts* in the spring of 1984 offered a profile of this constituency. ICR sent out five thousand questionnaires and received five hundred back, a respectable return for any mail questionnaire. ICR reported that 90 percent of its respondents are high school graduates, 73 percent have four-year college degrees, 35 percent have master's degrees, and 8 percent have doctorates. "This cross-section," it concluded, "surely refutes the common stereotype of creationists as ignorant and uninformed" (*Acts & Facts,* June 1984). (It could also have mentioned that many *Impact* articles are so technical that they would bewilder the ignorant and uninformed. Reading these regularly requires serious concentration, and casual browsers probably cull themselves from this kind of company.) The survey's occupational profile featured people in

117

The National Creationist Movement

business first (21 percent), followed by pastors (19 percent), teachers (18 percent), and those in industry (17 percent). The survey asked about the usefulness of ICR's literature as "spiritual help." Only 2 percent of the sample reported that ICR information was "instrumental in leading me to Christ," although 22 percent said that it was "effective in helping me win others to Christ." It is reasonable to conclude that ICR's constituency felt that creationism, even in its most effective, best-organized form, had no more than a modest power to convert people to Christ, but that its ability to enhance or intensify religious sentiments already in place was much stronger. The statement that ICR ministries were "important to my spiritual growth as a Christian" drew 61 percent of the sample. I interpret these results to mean that ICR's materials are far more effective at consolidating creationist sentiment, by reassuring conservative Christian readers that science corroborates creationism, than at turning doubters into creationists. ICR hardly brainwashes whomever it reaches, as its opponents might fear. Like many kinds of proselytizers, it preaches mostly to those who are already converted, and its effect is more to sustain the beliefs of the converted than to change other peoples' convictions.

For a final insight into the work of the Institute for Creation Research, one should appreciate "The Tenets of Creationism," by Henry Morris, the third major ideological point of reference of scientific creationism. The principal assertion of this document, published in July 1980, is that one can validly distinguish between biblical creationism, that is, the acceptance of Genesis 1 to 11 in terms of inerrancy, and scientific creationism, which is the creationist knowledge deemed appropriate for public school science education. The two are obviously varieties of the same belief—"two sides of the same coin," says Morris—but the conceptual distinction is critical for two reasons. First, no court will allow any teacher to advocate pure biblical faith in a public school science course. Second, Morris is concerned about the dangers of requiring nonbelievers to teach biblical beliefs: they might ridicule the Bible, expose it to critical examination, or lead their students into doctrinal deviation. Biblical creationism belongs in Christian forums like

118

The Core of Scientific Creationism

Sunday school classes, says Morris, while scientific creationism is the appropriate version for the public schools. Thus, biblical creationism assumes that "earth history [was] dominated by the great flood of Noah's day"; but, because this particular point is so clearly Biblical, the public school version would not teach "the Noachian Flood," per se, but rather "the worldwide evidences of recent catastrophism" (Morris 1980).

After stating those cautions, "The Tenets" lists nine beliefs each for biblical and scientific creationism. The tenets of biblical creationism are a condensed version of Morris's 1970 Christian Heritage College doctrinal statement: the Bible is inerrant with regard to creation ex nihilo, the six literal days of creation, Adam and Eve as the first humans, Noah's Flood, and so on; humankind is alienated from God by sin, but salvation through Jesus Christ atones for this if the sinner accepts Jesus; Satan, meanwhile, continues his rebellion against God; sometime in the future God will complete his plan of divine history; the home, the government, and the church are divinely ordained institutions that "should be honored and supported as such." The tenets of scientific creationism are, in effect, translations of that theology into a scientific vocabulary, following a four-part structure: they name a phenomenon of nature, state that the phenomenon did not evolve, then assert (in the passive voice) that it was created or was caused, and finally say that it was done by the Creator. Thus:

> The physical universe . . . has not always existed, but . . . was supernaturally created by a transcendant personal God . . .;

> The phenomenon of biological life did not develop by natural processes from inanimate systems but was specially and supernaturally created by the Creator;

> Each of the major kinds of plants and animals was created functionally complete from the beginning and did not evolve. . .;

119

The first human beings did not evolve from an animal ancestry, but were specially created in fully human form from the start. (Morris 1980)

Because these statements are not literal biblical passages, the creationists of ICR feel that they are excused from the accusation that scientific creationism is simple Bible study.

"The Tenets of Creationism" includes a third version of creationism called "scientific Biblical creationism," which is defined as "full reliance on *Biblical revelation* but *also* using *scientific data* to support and develop the creation model" (Morris 1980 [original emphases]). Morris did not elaborate this third form as much as the other two, which is unfortunate because this synthesis is the very heart of modern creationist belief. It is true that scientific creationism is not simply Bible study or Bible reading, but neither is it just a collection of objective technical facts unrelated to conservative theology. The fabric of scientific creationist belief is an interweaving of specific biblical beliefs with particular technical evidence, so that, for the creationist, the two explain and reinforce each other. Indeed, if scientific data can develop "the creation model" as Morris says, then the plenary authority of science rivals that of biblical revelation.

Morris's faculty statement of 1970 had elaborated the four plain points of CRS's 1963 pledge; this third document, "The Tenets," clarified some details of creationist belief from the 1970 statement. These periodic declarations are extraordinarily important to the creationist movement; they meet urgent demands for lucid standards that proclaim what to believe and what to reject. This may be difficult for evolutionists to appreciate, because the existential postulates of the two camps, creationists and evolutionists, are so drastically different. Not only do their basic documents differ, but even their needs for such statements are quite dissimilar. Creationists assume that, although human affairs are somewhat unstable and hard to understand, we nevertheless live within the larger, stable framework of God's unchanging reality. Consequently, creationists require statements to be produced occasionally to document that unchanging reality. The three documents between 1963 and 1980 do not reflect any change of beliefs, at least as creationists see them. Instead, they represent an increasing precision or specificity in describing the details of God's world.

Evolutionists, by contrast, are comfortable with the assumption that things change, even that they must change, as external circumstances change. Although it is not a formal postulate of evolutionary thought to define change as progress, there is still an implicit supposition that change is natural, so that unchanging beliefs, like other unchanging kinds of things, are vestigial at best and doomed to extinction at worst. Although many evolutionary authorities make many authoritative statements about evolution, none of those people and none of their pronouncements are considered timeless. Even Darwin's *Origin of Species* is, in many ways, a document specific to a particular historical circumstance, making it more a point of departure for the evolution of evolutionary thought than a gospel of categorical evolutionary beliefs. Evolutionists expect to find few, if any, unchanging points of ideological reference as they travel through life, and so they are often perplexed to see the power of the rigidly precise documents that specify reality for the creationists. Just as some creationists overestimate the influence of Darwinism by thinking mistakenly that evolutionists have elevated Darwin's book to the status of sacred scripture, so evolutionists often underestimate the declarations of Henry Morris that guard God's timeless realm in creationist belief. These misunderstandings grow as each party attributes its own existential assumptions to the other without realizing how irrelevant they are in the opposite contexts.

For noncreationists to appreciate the leadership of ICR in hammering out creationist belief under the direction of Henry Morris, it is important to realize that the rank and file of the creationist movement want a central authority that will clarify what to believe. Morris and ICR utter their words of absolutist belief in tones of rigid dogmatism, but the intent is not to force their views on people who disagree with them. Instead, they are serving a well-defined crowd—the consumers of creationist doctrine, so to speak—that demands the material ICR produces so well.

The National Creationist Movement

Institutions and Ideologies

The three core creationist organizations differ among themselves in methods, audiences, and goals. The Creation Research Society sees itself as a scholarly consortium that preserves scientific respectability for creationism by requiring its voting members to have bona fide advanced degrees in scientific or technical fields; it also guards against ideological deviation by requiring all its members to subscribe to a rigid doctrinal statement.

The Institute for Creation Research complements the work of CRS by projecting scientific creationism's moral theory and philosophy of science into mainstream U.S. culture. Its methods include lectures, seminars, debates, summer schools, publishing, and teaching. All those activities make ICR the best-known and most influential of the creationist organizations. Its staff members write most of the movement's seminal literature; they lead the movement in defining theological, scientific, and educational issues; and their monthly mailings to about seventy-five thousand subscribers reach more people than the mailing lists of CRS and CSRC combined. Thus ICR bears the main burden of teaching creationism to hundreds of thousands of conservative Christians, and of challenging the intellectual hegemony of evolutionary thought in the nation's secular universities. For these reasons, ICR is the engine that drives the creationist machinery.

The Creation-Science Research Center generally concurs with CRS's and ICR's moral theory and philosophy of science, but its strategy for projecting scientific creationism into U.S. culture is far more aggressive than ICR's. While it agrees that science is on the side of creationism, it is not much concerned about protecting creationism's scientific respectability. Instead, it attacks the teaching of evolution by means of a free exercise theory that alleges that *not* teaching creationism violates the constitutional rights of the Christian child. The logical consequence, however, is that creationism is equated with religion, not with science.

The ideological differences between CSRC and the ICR-CRS axis demonstrate that coming to terms with scientific au-

122

thority is not a simple matter, especially when that authority must be correlated with scriptural authority. Too much respect for scientific authority encroaches on scripture and subverts it, but too little runs the risk of being denounced for being unscientific, or even antiscientific. This problem becomes even more subtle, more varied, and more difficult when we consider two more creationist stances on scientific authority, namely, those of the Bible-Science Association and the Geoscience Research Institute.

Seven

Other Creationist Stances on Scientific Authority

The Bible-Science Association, whose leaders and constituents are mostly members of the Missouri Synod faction of Lutheranism, subjects everything about science to one simple test: does it conform with the fundamentalist understanding of biblical authority? BSA eschews the view that science is an independent authority that coexists alongside biblical authority. Instead, it supposes that information is scientific when it validates biblical authority *because* it validates biblical authority. The Geoscience Research Institute serves the Seventh-day Adventist church by making information about origins conform to Adventist dogma; yet it also injects a flavor of scientific skepticism into creationist thought by criticizing certain beliefs of the Institute for Creation Research on the grounds that those beliefs lack scientific merit.

The Bible-Science Association

The hallmark of the Bible-Science Association has been an authoritarian defense of the view that every bit of scientific information can be construed to prove that the Holy Bible is inerrant, with the result that scientific authority and Biblical authority are the same substance. BSA has a strong denominational character in that its officers are leaders of the Lutheran church, Missouri Synod, one of the most conservative factions of Lutheranism in the country, and their speaking venues are usually Lutheran churches. Walter Lang, a Missouri Synod preacher since 1940, founded the *Bible-Science Newsletter* in 1963. He was seeking technical advice from scientists about "Bible-

Other Creationist Stances

Science relationships"; recognizing that there was no satisfactory periodical providing such counsel, he established the *Bible-Science Newsletter* as a clearinghouse for this kind of information. Thus the newsletter, which reaches about seven thousand subscribers, is the oldest extant periodical of the scientific creationist movement (Lang 1983).

Lang established the Bible-Science Association in 1964 as a nonprofessional complement to the Creation Research Society; CRS would provide research and scholarship, while BSA would bring the message of scientific creationism to the public (Lang 1983; Numbers 1982). Although ICR has since eclipsed BSA in this role, BSA is still active and energetic. In 1972 it began to publish a monthly tract titled *Five Minutes with the Bible and Science* and eventually incorporated it into the newsletter. BSA has sponsored a national conference on creationism about every other year, and it organizes speaking tours for its officers. The BSA speakers' schedules for May 1983 through April 1984 listed fifty Lutheran venues, most of which were midwestern churches, plus twenty seven non-Lutheran host institutions (*Bible-Science Newsletter*, May 1983–April 1984).

When I interviewed Lang on 16 August 1985 at the BSA's conference in Independence, Ohio, he commented on BSA's relation to Missouri Synod Lutheranism:

> We know the differences thoroughly between the churches. They're not resolved. As we get more into conservatism, they're going to be stronger than ever. . . . I think the solution is not to try to develop more independent-type organizations, but the ultimate solution is for each church body to have their own active creation organizations. In 1970 I told [Henry Morris], I hope you make this [Christian Heritage College] strictly Baptist. He didn't appreciate that. I said, it's clear you're doing that, because you're with a Baptist church, and you're a Baptist, and it'd be much nicer if you'd just call it a Baptist school. At that time it wasn't particularly advisable for him even to do that, but that's what it's turned out to be. Everybody knows it's primarily a Baptist school. There's no problem

125

that everybody knows that we're primarily Lutheran. We both have been working in all kinds of different churches. . . . But it's awkward. It isn't the natural thing.

Other creationist groups have acknowledged BSA's special access to Lutheran groups. In our 6 December 1983 interview, Duane Gish of ICR noted that Lang "is readily accepted into the pulpits of Missouri Synod churches . . . so he has more of an impact there than he would have, say, in other denominations." Nell Segraves of CSRC mentioned when I interviewed her on 5 December 1983 that BSA is open to all denominations, but that it had an especially strong Lutheran membership. In her view, BSA began as a response to the problem of evolution infiltrating Lutheran seminaries; conservative Lutherans organized a series of seminars in their churches to deal with this problem, and BSA grew out of those seminars.

The Bible-Science Association took the long view in its plan to educate the public about creationism. After estimating that evolutionary thought took a century to seize control of Western science, the BSA leaders anticipated a hundred-year struggle of their own to recapture science for creationism. They said that the first twenty years of their program, from 1963 to 1983, was only the foundation phase of this effort (Lang 1983:10). This grand view of the future gave BSA the confidence to grow boldly even when its means were modest. When it found itself financially overextended in 1983, BSA interpreted that to mean that "the Lord has used situations like this to push us into growth-steps which we might otherwise be afraid to take" (BSA cover letter for mailing of September 1983). Furthermore, the BSA leaders have been able to construe their grand plan, and their roles in it, in biblical idioms. Lang, referring to himself in the third person, explained his own modest ministry this way: "Lang regards his work as being similar to that of a Samuel, rather than that of a David. Samuel was the educator, laying the foundation with the schools of the prophets which he founded. Later came the glory during the time of David and Solomon" (Lang 1983:10).

BSA's stance on scientific authority is to interpret every

issue, every question, every supposition, and every fact to prove every time that the Bible is inerrant: "We accept the Bible as the inerrant word of God, true in every subject which it touches, whether the plan be salvation, science, or history. Our acceptance of this position does not depend on what the methodology of science, history, or our current state of knowledge (or lack of it) may say about it" (*Bible-Science Newsletter*, April 1984:4).

BSA has three styles of exegesis: concordance, extrapolation, and free association. The first, exegesis by concordance, is a common way for conservative Christians to harmonize everyday reality with the Holy Bible. A person notices some new or interesting fact, then searches in a biblical concordance to find a passage that seems to have anticipated that fact, which then indicates that the Bible is true, and also that biblical truth predates modern discoveries by thousands of years. Consider one such exercise from the BSA's series, *Five Minutes with the Bible and Science*. Ancient humans lived in caves. This simple cliché seems to support evolution by referring to Peking Man, the Neanderthals of the Carmel Caves, Upper Paleolithic cave paintings, and so forth. The challenge to the Bible-Science Association is not to deny that cave dwellers existed, but rather to reinterpret the idea of cave dwellers in a biblical context. There are no cave dwellers per se in the Bible, but there are thirty nine references to *cave, caves,* and *cave's* in *Strong's Concordance*. Lot dwelt in a cave with his daughters (Gen. 19:30); the five kings who opposed Joshua at Gibeon hid in a cave (Josh. 10:16); Job's old enemies "dwell in the clifts [*sic*] of the valleys, in caves of the earth, and in the rocks" (Job 30:6). The Bible-Science Association then has a biblical grounding for cave dwellers. "Evolutionists believe that cavemen are ancestral to modern man, but Christians think of caves in connection with God's blessings upon the lowly" (*Five Minutes*, March 1975:13).

The next style, exegesis by extrapolation, does not require the same precise references. Its method is to state a belief and then declare post hoc that the Bible had inspired it, thus extrapolating loosely from the Bible according to one's intuition of biblical meaning. This gives the BSA exegetes considerable freedom to embrace UFOs, geocentrism, and other fringe beliefs that the Institute for Creation Research avoids. Some people

The National Creationist Movement

believe that UFOs are no less real than the Bible, and they expect an explanation of them in biblical terms. "In today's society," says BSA, "talking about dinosaurs, Noah's Ark, the Grand Canyon, fossil men, UFOs, etc., can lead to a connection with the Bible and its central message—salvation through Jesus Christ" (Lang 1983:10). BSA suggests modestly that "we believe that the Lord would encourage research of UFOs from a Christian viewpoint," which, it explains, is the notion that UFOs are agents of Satan who visit the earth to confuse and deceive us (*Five Minutes*, February 1975:3). BSA devoted the February 1975 issue of *Five Minutes with the Bible and Science* to the topic of UFOs, suggesting that UFOs were manifestations of the devil. "Creationists," it claimed, "have no need for life in space except for angels and devils." The genius of this explanation is that it saves a person from having to choose between believing in UFOs or believing in the Holy Bible. You can have both. UFOs are real, just as real as Satan, but a Christian should ignore them because they are satanic.

BSA's flirtation with geocentrism, the idea that the universe revolves around the earth, is another case of exegesis by extrapolation. Four creationist speakers explicitly advocated geocentrism at the 1984 Bible-Science Conference in Parma, Ohio, and one of them, Gerardus Bouw of Baldwin-Wallace College, said, "I would not be a geocentrist if it were not for the Scriptures" (*Creation-Evolution Newsletter*, May–June 1984:15). Geocentrism was even more prominent at the 1985 Bible-Science Conference in Independence, Ohio. Richard Elmendorf, a retired engineer from Pennsylvania, spoke for three hours in support of that theory, and he illustrated his talk with a spinning top, a pendulum, a turntable, a ball bearing, a frisbee, a gyroscope, a Slinky, and soap bubbles. Mr. Elmendorf said that the Bible implied that the earth is the fixed center of the universe, as in Job 38:4 (God "laid the foundations of the earth"). The Bible never says explicitly that the sun revolves around the earth, but Elmendorf extrapolated from verses that, with a little stretching, might be construed to imply as much, among them Josh. 10:13 ("The sun stood still in the midst of heaven"); 1 Chron. 16:30 ("The world also shall be stable, that it not be moved"); and Pss. 93:1 and 96:10 ("The world is stablished

128

Other Creationist Stances

[*sic*], that it cannot be moved"). In the evening of the last day of the conference, Bouw and James Hanson, of Cleveland State University, advocated geocentrism in a debate against two men defending the more conventional view, that the earth moves around the sun. Among other things, Hanson declared that "the Bible is overtly geocentric."

It was particularly interesting to see how solemnly the BSA audience behaved at these sessions on geocentrism. There was almost no criticism, or even discussion, from the audience. The only challenges came from several well-known anticreationists who attended. When I discussed the presentations on geocentrism with BSA members afterward, they commented, "This is very interesting"; "This is very impressive"; "I never knew that before"; and "There's a lot to learn about this." Not one of them rejected geocentrism.

Finally, there is a line of BSA exegesis even more tenuous than concordance or extrapolation. I call this exegesis by free association, for BSA statements in support of creationism that have no apparent grounding in any commonly accepted exegesis, apologetics, or theology. Three examples illustrate this style:

(1.) Creationist ecology is superior to evolutionary ecology in reviving deoxygenated lakes. "Many governmental agencies . . . have had little success because they have been working with an evolutionary world view. In contrast to this, Bob Laing of Clean-Flo Laboratories, Inc., in Minneapolis, has successfully restored many dead and dying lakes to full life, working from a creationist point of view" (*Five Minutes*, May 1982:3).

(2.) Creationist physics is replacing Einsteinian physics. Creationist Thomas Barnes "has solved some of the problems which Einstein was able to see, but never resolve. . . . Many creationists believe that . . . [Barnes's] work will eventually see Einstein replaced by Barnes in the world of physics" (*Bible-Science Newsletter*, October 1983:12). (Barnes, who is affiliated with ICR, is considerably more modest than BSA in his claims about his own work.)

(3.) BSA is the only major creationist group in the country that gives serious attention to the work of Barry Setterfield, an Australian creationist who rejects the basic assumption in physics that the speed of light is constant. The speed of light (186,000 miles per second), when combined with astronomy's measures of interstellar distances, provides estimates for the age of the universe. If a distant star is fifteen billion light-years away, and the speed of light is constant, then the universe must be at least fifteen billion years old. But if the speed of light was greater in the past, then light from that distant star would have reached the earth sooner, thus compressing the age of the universe to fit into a creationist chronology. Setterfield proposes that the speed of light was much greater in the past, and that it approached infinity six thousand years ago, which, coincidentally, is the approximate date of creation in Archbishop Ussher's chronology. He also concludes that the speed of light stabilized at its present rate in about 1960. Thus a star could still be a *distance* of eighteen billion light-years away without requiring a *time* of eighteen billion years for its light to reach Adam and Eve on earth. Most creationists disdain this daring challenge to the most basic constant in physics, but the *Bible-Science Newsletter* endorses Setterfield enthusiastically (July:5ff., October 1983:12).

There are powerful ironies in BSA's exegesis. If everything scientific has to be harmonized with the Bible one way or another, then there is no logic that is unique to science. In these terms, science is not a particular way of thinking, or even an organized body of knowledge. It is only a corpus of exegetical evidence. Whatever can be construed to illuminate scripture is scientifically valid. All this can be comforting to Bible-believing Christians who accept UFOs or geocentrism. For others, however, it raises disturbing questions about the Bible and its message. A potential convert might wonder whether all the topics in BSA's ideology are integral to biblical belief. Does becoming a Christian require accepting Elmendorf's geocentrism, or Setterfield's speed-of-light curve? And a Christian who is comfortable with the Bible might ask whether scripture needs this kind of information in support of its moral message. Do cave dwellers, whether creationist or evolutionist, enhance Job's long lyrical sermon on coming to terms with suffering? Do the Psalms lose their meaning if the earth really orbits the sun?

BSA's judgment on scientific matters distresses some creationist leaders. One told me confidentially that he had repeatedly protested to Walter Lang about unscientific statements that had hurt the cause of creationism, but without apparent effect on Lang's ministry. Still, the structural position of the Bible-Science Association within the creationist movement is all but inevitable. A movement claiming to adhere to inerrant scriptural authority must have a faction that appoints itself to guard that source against all critics. If the Bible-Science Association was not doing this, then the responsibility would devolve to ICR or CRS to do the work of compelling every item of science to affirm the authority of the Bible. In a sense, BSA is only taking the CRS-ICR philosophy to its logical extreme.

The Geoscience Research Institute

The Seventh-day Adventist Church sponsors the Geoscience Research Institute to screen scientific information about origins so that Adventists can reject information that conflicts with church dogma. This denomination is deeply committed to the idea of creation week (that Genesis 1 describes six literal twenty-four-hour days) as well as most of the other principal decrees of the Institute for Creation Research. Likewise, it is committed to most of the major features of fundamentalist thought, including the plenary inspiration of the Holy Bible. Seventh-day Adventists maintain a distinct identity among fundamentalists, however, by virtue of their Saturday Sabbath and the importance they attribute to the interpretive prophecies of their nineteenth-century founder, Ellen G. White. As the Adventists are unique among fundamentalists, so their creationist organization, GRI, is unique among creationists. The theological chart they use to navigate the issue of origins is quite independent of any other creationist program.

Ellen G. White established the Seventh-day Adventists' opposition to evolution in 1864, barely five years after the publication of Darwin's *Origin of Species*. She wrote: "When men leave the word of God in regard to the history of creation, and seek to account for God's creative works upon natural principles, they are upon a boundless ocean of uncertainty." And "the genealogy of our race, as given by inspiration, traces back its origins, not to a line of developing germs, mollusks, and quadrupeds, but to the great Creator" (Review and Herald 1976:443).

Whereas most of the early conservative Protestant opposition to evolution focused on the biological issue of human origins, the Adventists concerned themselves more with geology and geochronology because the church harbored "a deep suspicion of human reason, and nothing seemed to confirm this suspicion better than the science of geology, which depended so crucially on the *assumption* of uniformity" (Numbers 1979). Thus the geologists' inferences and extrapolations about events in the long-ago past, which could never be verified, appeared to the Adventists to be little more than a thin tissue of guesses and conceits.

Their other objection to evolutionary geology centered on the very meaning of the denomination's name. The Adventists celebrate the Sabbath on Saturday, for this is the seventh day of the week, the day God rested after completing the creation. If the first chapter of Genesis represents allegorical time periods, as opposed to seven literal twenty-four-hour days, then the Saturday Sabbath and the "Seventh-day" name become irrelevant. Ellen G. White and her associates confronted this problem and solved it by declaring that the creation week comprised seven literal days (Numbers 1979:20). More recently, the church has proclaimed that "the Sabbath as an institution [is] inseparable from the Creation," so that the celebration of the Sabbath on Saturday is the church's "memorial of creation" (Review and Herald 1976:357, 1957:12). On 1 December 1983 I interviewed a group of GRI's scientific staff members in Loma Linda, California: Robert H. Brown, Ph.D. (University of Washington); Harold Coffin, Ph.D. (University of Southern California); Ariel Roth, Ph.D. (University of Michigan); Richard Tkachùk, Ph.D. (University of California, Los Angeles); and Clyde Webster, Ph.D. (Florida State University). According to Coffin, this conviction of the Sabbath as memorial "makes this matter of literal creation week, and creation in general, a very important aspect" of Adventist thought. The first edition of GRI's journal, *Origins*, included an editorial reminder that the "seventh day" of the denomination's name "refers to the Sabbath, the memorial of God's creation, a weekly celebration of His creatorship" (Neufeld 1974). In short, the tradition of antirationalism plus the defense of the meaning of the denomination's name caused Adventists to see geological issues as theological questions (Numbers 1979:20).

Occasionally the church heard a different opinion. In 1874 and again in 1879, John Harvey Kellog, a teacher at an Adventist college, urged his coreligionists to concede some independent credibility to scientific findings, and so to avoid pointless conflicts between faith and science. But opinions like his were eventually overcome by the leadership of George McCready Price, an amateur armchair geologist who fine-tuned the Adventist hostility to geological reasoning. The theme of Price's long career of teaching at Adventist schools and writing was his own "great law of conformable stratigraphic sequences": "any kind of fossiliferous beds whatever, 'young' or 'old,' may be found occurring conformably on any other fossiliferous beds, 'older' or 'younger' " (Numbers 1979:18–23). In other words, geology makes no sense with regard to earth history. This was utterly specious, in the judgment of the scientific community, but it was nevertheless useful to Adventists and other creationists who desired an earth history unencumbered by geological insights.

Price's views on the geological record made it easy for him to insist on the Ussher chronology, according to which the creation occurred in the year 4004 B.C. Any deviations from that timetable, including gap theory and day-age theory, were "the devil's counterfeit" and "theories of Satanic origin" (Numbers 1979, 1982).

Price and other Adventists formed the Deluge Geology Society in 1938 with the intention of promoting their narrow creationist orthodoxy (Numbers 1979:26–27). At the same time, however, two of Price's former students who had since acquired their own scientific expertise, Harold W. Clark and Frank Lewis Marsh, began to disagree publicly with Price on matters of geology and biology. The church's leaders recognized a pressing need for well-educated staff scientists who could guide Adventists past the double-edged difficulties of geology-theology disputes, so they began to send reputable believers to graduate school at secular universities. As GRI staff members reported in our December 1983 interviewing, by the mid-1950s Price's society was unable to satisfy Adventists' needs for information about evolution and creation, for science teachers at Adventist schools were circumventing Price by appealing directly to church headquarters for better information. The church responded by gathering its newly minted Ph.D.'s into the Geoscience Research Institute, founded in 1957, "for the purpose of making available to the SDA Church competent advice concerning relationships between the natural sciences and inspired testimony" (Roth 1974).

Although the Geoscience Research Institute seemed poised to deliver the denomination from narrow orthodoxy in scientific matters, it soon became an new instrument of old dogmatism. Despite his disagreements with George Price, Frank Marsh, first director of the institute, took the view that science should be closely guided by the teachings of Ellen G. White. In 1964 a colleague who thought otherwise found himself separated from GRI. Similarly, a 1971 purge removed others who might have deviated from church teachings. "The church's brief experiment with 'openmindedness' thus came to an end" (Numbers 1979:27).

The Geoscience Research Institute commits itself primarily to serving the needs of science teachers in the Adventists' parochial schools. It provides an intradenominational service, rather than an evangelical outreach to the public. Consequently, it is hardly concerned about the central political dynamic of the evolution-creation controversy, which is what to teach in public school science courses. Harold Coffin and Ariel Roth of GRI testified in support of the scientific merits of creationism at the Arkansas trial in 1981, but their colleague, Robert Brown, told me in our interview that he endorsed Judge Overton's decision for preserving church-state separation. Clyde Webster and Richard Tkachùk, two other scientists at GRI, commented during the interview that public school issues like the Arkansas case were not particularly important to them when compared with their own denomination's parochial school concerns.

Another distinction between GRI's creationism and that of ICR and CRS is GRI's endorsement of gap theory. In this theory, the creation "in the beginning" of Gen. 1:1 was a different event from the six-day creation starting in Gen. 1:3, with a long period in between during which "the earth was without form, and void" (Gen. 1:2). In the GRI version, life began with the very recent six-day creation described in Genesis, but long periods of geological change could have preceded that moment. Gap theory worries the strict creationists at the Institute for Creation Research because it offers a logic that could reconcile biblical creation with the long chronologies required by evolution (which is also why George McCready Price opposed it so bitterly). This certainly does not turn those biblical verses into evolutionary statements, but it nevertheless compromises the simple opposition of evolution versus the Bible. The GRI scientists told me in our joint interview in December 1983 that they realize their gap theory bothers the Institute for Creation Research. Nevertheless, they stick to their interpretation because they cannot accept that all of the earth's history, as manifest in the geological record, could fit into ICR's short chronology of about ten thousand years. Gap theory, say the GRI people, is not a theological postulate but a scientific conclusion.

Meanwhile, GRI adheres to creationist orthodoxy on another major issue of chronology, namely, day-age versus creation week. Day-age is the supposition that the six days of creation in Genesis 1 were long geological ages, while creation week holds that they were six literal twenty-four-hour days, plus one twenty-four-hour day of rest. The official Seventh-day Adventist position on this is clear: "We believe . . . that God created the world in six literal days; we do not believe that creation was accomplished by long aeons of evolutionary processes" (Review and Herald 1957:24). Naturally, GRI conforms to church doctrine on this issue. The consequence of combining gap theory with creation week is to endorse the process of divine creation, but without committing oneself to an implausible geochronology. The six days of Genesis 1 are said to be literal days, and they can be relatively recent events, but the bulk of the geological record can refer to events from millions of years ago, during the conditions mentioned in Gen. 1:1–2. (To ICR and the other strict creationists, this kind of chronology is nothing more than exegetical shenanigans.)

With creationist credentials secured by the history of their church, and scientific credentials as good as anyone else's, GRI frankly criticizes other creationists' judgment. It offers the most penetrating criticism within the movement, revealing that there is considerably more disagreement among creationists than many people realize, by trying to separate plausible creationist arguments from preposterous ones. Three kinds of poor judgment draw GRI's criticism: (1) the style and tone of attacks on evolutionists; (2) the lack of scientific standards in certain interpretations of empirical data; and (3) simplistic deductive arguments. These things bother the GRI staff because they feel that scientists need salvation as much as anyone else (Kootsey 1976), so that Christians in the scientific community must earn the professional respect of their colleagues if their gospel message is to be credible. They realize that two sure ways to squander that respect are to launch ad hominem attacks on evolutionists, and to present negative views on evolution in lieu of positive views on creation. Consequently, they warn that "we should carefully avoid statements that ridicule evolutionists, make them look foolish, dumb, uninformed, or in any way tend to make them look bad" (Brand 1979). Likewise, they recommend avoiding "the negative approach . . . [that is,] a recitation of the errors made by evolution-oriented scientists, . . . a list of hoaxes perpetrated in the name of evolutionary science" (Neufeld 1974).

Consider a few critical comments by GRI staff and other Seventh-day Adventist scientists. Robert Brown in our December 1983 interview said, "We feel quite unhappy about scientific arguments that are promoted widely to support the creationist view which we know do not stand scrutiny according to ordinary scientific principles of evidence gathering and interpretation. And it puts us in tension with some of these folks who we like to feel we have complete agreement with."

Leonard Brand, arguing for the integrity of the creation argument, complains that "much of the material that has been published in support of creation is full of errors and should not be used. Sometimes Christians feel a burden to defend creation and proceed to write books or journal articles on the scientific evidence for creation without having the scientific training to match their zeal" (Brand 1979).

The sloppy science that most disturbs GRI is the case of the Paluxy River footprints. There are geological deposits beside the Paluxy River in Glen Rose, Texas, that contain the tracks of three-toed dinosaurs. There are also depressions, in the same deposits, that many creationists have interpreted as human footprints. Until 1986, the creationists at ICR concluded from this that humans and dinosaurs lived at the same time, implying that evolutionary chronology had a fallacious gap of about fifty-five million years between the end of the dinosaurs and the beginning of the hominids. GRI rejected this interpretation of those geological features, and its criticism was merciless. Berney Neufeld wrote that

> the Glen Rose region of the Paluxy River does not provide good evidence for the past existence of giant men. Nor does it provide good evidence for the coexistence of such men (or of other large mammals) and the giant dinosaurs. . . . In any kind of investigation, but especially when investigating the past where data are more equivocal, caution and thoroughness should characterize the work done, and conclusions should not be drawn prematurely. (Neufeld 1975)

Robert Brown, who specializes in geochronology at GRI, commented at length on his own research on the Paluxy prints in our December 1983 interview:

> Our creationism is not based on just armchair work. We've done a number of field studies at the Paluxy River on which we base our reticence to promote that as a viable argument. . . . I've been down there on at least four or five occasions with a group of scientists, and there's nothing there that I would have any confidence promoting as solid evidence. Everything I've seen there has a far better explanation. . . . I first went down there with a group of scientists *wanting* to see it, and said, look, the guys who have told me that that

isn't real human footprints, they are liberals trying to brainwash me and undermine my religious experience; I'm going to take a fresh look and go down there and see the evidence. Well, I hadn't been there one day when I came away and said, no way can I go to any person and responsibly call this evidence "looking for footprints."

GRI justifies its skepticism by turning to the epistemological principles of Ellen G. White, which emphasize inductive thinking more than deductive. For example, Robert Brown outlines the inductive-deductive distinction in an *Origins* editorial criticizing an article from the *Creation Research Society Quarterly*:

> The salient idea in each of these statements [in *CRSQ*] is derived from the Bible and is not a natural product of scientific observation or inductive reasoning from such observation. . . . Neutral Scientific Creationism operates independent of religious concepts and traditions. It may involve hypothesis and deduction, but it places principal emphasis on inductive logic. . . . Creationism that derives its basic ideas from the Hebrew-Christian scriptures, and then uses science to further develop those ideas, is Biblical creationism, a subcategory of Apologetic Scientific Creationism. In contrast with Neutral Scientific Creationism, the emphasis of Apologetic Scientific Creationism is on deductive reasoning. It begins with a religion-based theory. (Brown 1981)

The subtlety of this distinction might well escape most critics, for whom all creationism is Biblically-deductive, leaving only a distinction between honest deduction and veiled deduction. But the people at GRI are not splitting hairs for the sake of creationism's critics. They are striving to realize the kind of thinking posited by Ellen G. White, in which the Bible is a great source of truth, of course, but not the only source. Experience is another, no less legitimate than the Bible, and inductive em-

138

piricism is part of experiential truth. Certainly there is no desire to devalue the Bible in the expansive epistemology of the Seventh-day Adventists, but neither is there a demand to constrain all truth within biblical idioms, as in the mechanical exegesis of the Bible-Science Association. It shocks the core of the creationist movement to hear that some sources of truth might stand independently of the Bible, but GRI is more than comfortable with Ellen G. White's legacy.

Intellectual danger lurks in this logic: What happens in a conflict between two equally legitimate forms of truth? What does the Adventist do when truth from empirical science differs from biblical truth? There is an answer that anticipates this problem and frustrates the stern exegetes inside creationism as well as the sardonic critics outside, all of whom might prefer to watch the Adventists contradict themselves with their offbeat epistemology. Metaphysical holism is the solution that, for the Adventists, dissolves the dilemma. "Adventists are holistic," said Richard Tkachuk of GRI in our 1983 interview. "We do not dichotomize our knowledge into science and religion, but view life as a whole." If all truth, whether scientific, scriptural, or experiential, has the same undifferentiated fabric, then it is not possible to compare the pattern of one form of truth against that of another. In order to subsume science under religion, in the style of the Bible-Science Association, or to set religion against science, in the strategy of creationism's worst enemies, it is necessary to break the whole truth down into distinct analytical isolates—science here, religion there. This is just what the Seventh-day Adventists at GRI refuse to do. Their truth is amorphous, resisting clear internal divisions, and consequently compatible with almost any external source of truth. This naturally leaves the GRI scientists unable to connect science to religion with any precision or specificity. They are competent philosophers of science and articulate Christian witnesses, but their perception of how the two spheres connect is not more complicated than Ariel Roth's comment during our 1983 interview that "truth should integrate itself as a whole."

Although it cannot resolve this problem of relating science to religion, the Geoscience Research Institute devotes considerable attention to it. Many articles in *Origins* refer to this topic, but they generally come to terms with it by advocating broad definitions of science and religion, thereby avoiding conflict. I do not believe that the scientists writing in *Origins* are intentionally evasive, or that their views are specious. They see only a nonissue where most other principals in the creation-evolution controversy see a major substantive question. Adventist ambiguity in this matter is a problem only for non-Adventists.

The journal *Origins* is GRI's principal vehicle for disseminating its views. The introductory editorial in its first issue explains its purpose by referring to GRI's responsibility to science educators in church schools. Its editorial consultants have affiliations with such Adventist institutions as Andrews University, Loma Linda University (where GRI is now located), and the church's general conference. Most of the authors teach at church schools. Many articles refer to the writings of Ellen G. White to support their conclusions. At the same time, however, the tone of many articles is well within the secular mainstream of philosophy of science. For example, Gary L. Schoepflin's brisk survey of scientific paradigms, "Perceptions of the Nature of Science and Christian Stategies for a Science of Nature," begins and ends with theological concerns, yet the substance in between is a useful guide for any scientist, regardless of religious affiliation (Schoepflin 1982).

GRI's relation to the creationist movement is a curious combination of internal authoritarianism and external skepticism. Within its own church, it conforms to the well-established Seventh-day Adventist tradition that holds creation prominent in the church's theology, and even in its name. This emphasis helps the denomination retain a distinctive identity within the bewildering kaleidoscope of conservative Christian groups in the United States; the Seventh-day Adventists show by their name that they are as creationist as anyone else. A long tradition of anti-evolutionism, stemming from their founder, reinforces this sentiment. GRI is proudly faithful to the anti-evolutionary values of its parent denomination.

Then there is GRI's external role, according to which it directs skeptical judgments and harsh criticisms against ICR's scientists and BSA's mechanical exegetes, thereby shearing itself away from the ideological postures of the mainstream creationists. Actually, the GRI scientists focus their disputations largely on empirical particulars and isolated deductive interpretations, so that they never bring their skepticism to bear against the broad suppositions that give creationism its religious framework. They argue about the trees but take the forest for granted, so to speak. This way, GRI is always generally creationist in its outlook, but often specifically anticreationist in its individual articles about particular issues.

In spite of the criticisms it dishes out so freely, GRI seems to get along well with most other creationist groups. Not much creationist literature complains about the views of the GRI scientists or even responds to them. When I asked leaders at ICR and CSRC about differences between themselves and the Geoscience Research Institute, they spoke kindly of GRI. Robert Brown of GRI reminded me in our interview that "we do everything we can to maintain cordial Christian relationships with [ICR] and to give them support and encouragement . . . We respect [ICR] for their commitment and for the giant efforts they have made."

Coda: Moral Theory and Scientific Authority

When conservatives refer to the creation stories of Genesis as the basis for their moral theories, they presume that the historical authenticity of those stories, and the moral implications thereof, are eternal; similarly, scientists assume that the laws of nature are timeless. If scientific creationism is both scientific and creationist, then it must be a timeless body of truth, according to both criteria. And yet the rise of scientific creationism is very much an artifact of certain conditions in late twentieth-century America.

The first of these conditions is the rise of the plenary authority of science. Rightly or wrongly, science is widely believed to be the key to knowing what reality is, not only for tangible matters like medicine, energy, and consumer technology, but also for less certain concerns that are loaded with moral import, such as human behavior and social policy. The secular grace of scientific sanctification enhances almost every kind of claim in almost every sort of issue. Strangely, however, it is not the intellectual structure of scientific thinking that makes a claim credible. Neither organized skepticism, nor naturalistic explanation, nor other patterns of scientific reasoning are widely appreciated, let alone persuasive. It is not the substantive content of scientific knowledge, either, that earns respect for a given product, theory, or policy. The institution of science produces far too much empirical information for any one citizen to master. Even a person with a doctorate in science has to be selective about what he or she knows and cares about. Rather, it is the artful deployment of the easily recognizable symbols of science that convinces a citizen, a customer, or a client that product X, theory Y, or policy Z is the right one to choose, on the grounds that it has been blessed by scientific sanctification. Unfortunately, the twentieth-century process of popularizing science has fragmented the logical relation between the intellectual structure and the substantive content of science, on the one hand, and the common symbols of science, on the other. The result is that those symbols are available for trivial and superficial appeals. Scientific authority must be reckoned with, but it is possible to do so by borrowing its symbols while ignoring its intellectual structure and substance. Fundamentalist anti-evolutionists, like so many other interest groups, have to come to terms with the plenary authority of science, but they have considerable freedom to interpret and deploy that authority to suit their own purposes.

The other condition that has shaped modern creationism is the moral situation of U.S. culture in the postwar decades. While it is common for fundamentalists to feel that social change equals moral disintegration, that sentiment was especially acute in the 1950s and 1960s. The evangelical Protestant code of morality and virtue, which had set the terms for public culture for many generations, now had to share civic life and the public schools with other points of view. This is not to say that conservative Protestants were forbidden to honor their own morality and virtues, but they had to tolerate the coexistence of atheists, agnostics, Catholics, Amish, Jews, and others, on more or less equal terms. To those dissidents, this was freedom of religion, but to many a conservative Protestant, this was a nation turned upside down. (Meanwhile, liberals saw the same period as the age of Joseph McCarthy, J. Edgar Hoover, and Dwight D.

Eisenhower, that is, a time of conformism and cultural conservatism.) Especially painful were the great Supreme Court decisions on religion, from *Torcaso* in 1961 to *Lemon* in 1970. And for some, there was a suspicion, sharply felt but ill defined, that evolutionary thought had a hand in the moral disintegration of that age.

Arriving at the historic confluence of these two themes was the ministry of Henry Morris. Actually, John Whitcomb, Jr., had been working on the theology of creationism for a decade, and Morris had devoted almost two decades to the scientific basis of creationism, but the publication of their book *The Genesis Flood* was the spark that lit the tinder of fundamentalist anxieties about science and morality. Subsequently, scientific creationism came close to dismantling evolution's position in science education in some states between March 1961, the month of that book's appearance, and March 1981, when Arkansas passed its equal-time law. All along, creationism has defaced the popular image of evolution as an icon of modern science and has dissipated the fear once held by the country's fundamentalists that believing Bible stories required renouncing scientific authority.

Scientific creationism's most precious goal, to have its views included in public school science education, is unlikely to be achieved except in a limited number of local school districts. But in more general terms, scientific creationism has changed the nation's assumptions about the credibility of evolutionary thought and has given conservative Christians reason to believe that science is the Bible's best friend, so that they can merge their faith in sacred scripture with the science-worshipping habits of modern U.S. culture.

Modern creationism cannot be reduced to either scientific illiteracy or a slavish devotion to some verses of scripture. In fact it is a rich, complicated, and varied system of knowledge, values, and beliefs (cultural meanings, as the anthropologist says) that enable fundamentalist and evangelical Christians to come to terms with certain realities, anxieties, uncertainties, and changes in U.S. life. In particular, the two themes that constitute the existential backbone of contemporary creationism are: (1) How does one come to terms with the plenary authority of science, as compared with scriptural authority? and

(2) How does evolutionary thinking contribute to the social changes that conservative Christians diagnose as moral decay? The ultimate question is, How are these two matters related? For the apocalyptic separatists the answer is simple. Secular culture is a satanic landscape of peril and deceit; science is part of secular culture; evolution is part of science. Thus the two questions are neatly unified. (But according to that line of thinking, scientific creationism is also secular because of its scientific pretensions.)

For other kinds of creationists and potential creationists, these issues are not so simple. Within the broad realm of conservative Protestantism, the philosophical opposite of the separatists' stance is the troubled ambivalence of the American Scientific Affiliation (and others of the evangelical center), whose respect for biblical revelation inclines them to sympathize with creationism, but whose evangelical ethos and scientific judgment shy away from *scientific* creationism. A different kind of ambivalence is the creationism of the inerrancy Baptists, who hardly care whether science agrees or disagrees with their exegesis, since the word of God is perfectly credible in any circumstances.

In between the apocalyptic separatists and the American Scientific Affiliation is a cluster of organizations for whom it matters very much that science corroborates the historical authenticity of the narratives in Genesis. Foremost among them is the Institute for Creation Research, the most effective creationist organization in the land because it presents the most convincing program for linking biblical creationism to scientific credibility. Millions of conservative Christians ask to be reassured that their version of origins has some scientific sanctification. In truth, no other creationist organization can display scientific credentials, terms, or symbols as skillfully as does the ICR staff. Assisting in that function is the Creation Research Society, whose contribution is to buttress ICR's scientific image with its technical journal and its members' academic degrees.

The irony in this is that ICR's followers take science more seriously than most scientists do. Like so many Americans, Henry M. Morris and his colleagues at the Institute for Creation Research exaggerate the ability of science to sanctify the meanings we find in our lives. Their proposition, that science confirms timeless truths, is dubious to most of those who know how circuitous scientific discovery can be, and how tentative its conclusions are. But because this habit of exaggeration is so widespread in our nation, ICR's proposition resonates with many citizens.

On the left flank, so to speak, is the Geoscience Research Center, the unofficial skeptic of the creationist movement, which heartily embraces the general principle of scientific creationism, for theological reasons, but rejects many specifics of ICR's brand of creationism, for empirical reasons. The right flank is the Bible-Science Association, whose understanding of science corroborating scripture is that, if something corroborates scripture, then it must be scientific. Also there is the Creation-Science Research Center, which has a philosophy of science and scripture like ICR's yet pursues its mission through legal actions, not scientific arguments, with the result that its stand on science is moot.

Within that cluster of creationist organizations centered on ICR, it is universally agreed that the idea of evolution bears much responsibility for immorality. This the GRI discovers by studying the writings of Ellen G. White, founder of Adventism. The others come to that conclusion by means of the negation of personal belief theory about Secular Humanism, or the autonomy theory, or the view that evolution is directly responsible for immorality, or some combination of those explanations, each of which includes a denunciation of evolution.

And yet that broad moral theory does not connect very closely with the scientific creationist stance on the plenary authority of science. Essentially there are two separate arguments, one about morality, the other about science. Creationism is good, because the plenary authority of science authenticates it, and evolution is evil, because immorality taints it. The scientific creationists urgently want science to make sense morally according to both of those arguments. But even if the two complement each other, neither leads logically to the other. Either stands or falls on its own.

Because creationism includes an array of understandings about scientific authority, and a spectrum of moral interpretations of evolution, and especially because the two topics can be resolved separately, creationist thought is far from being a unified theory about how science makes sense morally. On the contrary, it is a kind of rows-and-columns matrix, with a row of choices on scientific authority, and a column of interpretations of morality. The converts and potential converts to the cause are confronted with a considerable selection of doctrines, attitudes, and positions, each of which is genuinely creationist. To some extent a believer's choices are narrowed down by sectarian preferences. Most Adventists prefer the GRI line, and conservative Lutherans likely embrace the BSA outlook. ICR, however, has a moral theory that is not distinctly Baptist but rather a generic fundamentalism, even if its social base is strongly Baptist. Thus modern creationist thought, because it is far from unified, poses choices. The individual creationist or potential creationist has to select some combination of science and morality from the array offered by the national creationist movement.

That question within a question is the subject of part 3. Again, the answers are to be found in social circumstances and existential considerations, but now those answers are rendered in much finer texture: the case of North Carolina placed under a microscope, as it were. So, within the sweeping national considerations of morality and scientific authority, there are local forces to consider. The parochial politics of religion and education within North Carolina narrow the creation-evolution controversy by requiring North Carolina creationists to make their claims in terms that can be clearly understood by their neighborhood churches, schools, and politicians.

Furthermore, the lives and feelings of individual creationists are different in interesting ways. Their own educations, occupations, and church affiliations color their attachments to creationism. In summary, part 3 asks how the broad considerations described in part 2 affect and are affected by the more intimate circumstances of individual creationists' lives, and by the social dynamics of church and school in North Carolina. By tying the concrete personal experiences of the local activists to the grand ideological themes of the creationist cause, we learn why creationism matters so much in some peoples' lives.

Part Three
Creationism in North Carolina

Eight
Fifty Years of Creationism and Evolution in North Carolina

North Carolina endured a long episode of anti-evolutionism in the 1920s. That experience, so common in the southern states, seems at first to suggest that evolutionary thought is chronically vulnerable to fundamentalist depredation, and that evolutionists must always be on the defensive. But this conclusion is superficial. The events of the 1920s and the years since reveal that the idea of evolution has been more durable, and creationists less intimidating, in North Carolina than in southern states such as Tennessee, Mississippi, and Arkansas, where evolutionists did indeed suffer serious setbacks.

Unlike Tennessee's experience of anti-evolutionism in the 1920s, which climaxed in a personal clash of the titan orators William Jennings Bryan and Clarence Darrow, the events in North Carolina occurred without much help or hindrance from national celebrities. Several nationally known fundamentalists passed through the state to make cameo appearances in its creation-evolution dispute, but local political and religious considerations determined the outcome. Several of the principals in the dispute were outstanding people well-known around the state, but none were as famous as Bryan or Darrow. Understanding the developments in North Carolina lets us see how old-time creationism affected ordinary people, and how they, in turn, affected those developments, particularly the campaign within the State Baptist Convention to purge evolution from Wake Forest College, and the battle in the state legislature to outlaw the teaching of evolution at the University of North Carolina (UNC).

149

Creationism in North Carolina

First the Baptists. In 1920, T. T. Martin, a rabid anti-evolutionist from Mississippi, aroused fundamentalist Baptists in North Carolina by charging that William Louis Poteat was a "rank infidel" because he taught evolution in his biology classes at Wake Forest College, which was owned by the Baptist State Convention. Poteat was both a great scientist and a great Baptist: a leading biologist, a past president of the North Carolina Academy of Science, a past president of the State Baptist Convention, and president of the college. Because Poteat taught evolution at a Baptist institution, Martin accused him of everything from undermining biblical faith, to thinking like a German militarist, to misappropriating funds (Gatewood 1966:30–34).

Poteat and his supporters were in a difficult position. The Baptist colleges in the state had just begun a great fund-raising campaign, but the fundamentalist forces threatened to withhold their contributions from schools that taught evolution. The charges against Poteat were sweeping, reckless, and ambiguous, with the fundamentalists holding evolution responsible for "secularism, heresy, sexual immorality, 'busy divorce courts,' the collapse of parental authority, adultery, juvenile delinquency, and the disintegration of the family," not to mention "infidelity, pessimism and suicide," plus atheism, Bolshevism, and German militarism (Gatewood 1966:98, 229–230). This was an early manifestation of the theory that the idea of evolution is directly responsible for every kind of immorality imaginable. The denomination had no dignified or impartial procedure by which Poteat could face his accusers or defend himself. The issue could only be settled in a resolution at the annual meeting of the State Baptist Convention, after wild fears and scurrilous smears had spread throughout many local Baptist associations. Even then, the decision would be a simple majority vote. Says historian Willard Gatewood:

> The peculiar organization of the Baptist denomination allowed those [fundamentalist] elements to fan the evolution fires in local congregations and associations. Try as they would, the Baptist leaders who desired to avoid a convention squabble found it increasingly difficult to soothe these local groups so

agitated by the disturbances over evolution. One observer noted that, in North Carolina, Baptists of extraordinary distinction had 'no more authority in matters of faith and doctrine than is possessed by some semiliterate pastor of some Little Bethel in the remote backwoods.' (Gatewood 1966:70)

Poteat faced his detractors at the state convention in December 1922. In a moving sermon, he presented his views on Baptist education. "The central theme of his address was the need for an enlightened faith spacious enough to accommodate all realms of culture"; Christian schools "must acquire and utilize all the knowledge that science offered." Though he insisted on the right of professors at Baptist colleges to teach evolution, he spoke so convincingly as a devout Christian that his enemies were utterly disarmed, even to the point of endorsing a resolution that praised Poteat and his views (ibid.:74–75; Linder 1963).

This was a personal vindication for Poteat, but it should be seen in the larger political context. The most important faction at that meeting was a large centrist group that probably disagreed with Poteat about evolution but abhorred the intradenominational strife perpetrated by the fundamentalist wing (Gatewood 1966:73). Guiding the centrists were Baptist officials Richard T. Vann, president of Meredith College; Livingston Johnson, editor of the *Biblical Recorder*; and his brother Archibald Johnson, editor of another important Baptist publication. Historian Tom Parramore refers to those three and likeminded Baptists as "enlightened fundamentalists," meaning that they were biblical literalists who personally rejected Darwinism but nevertheless publicly defended academic freedom for Poteat and the denomination's other college teachers (Parramore 1976). Theirs was a vote for academic freedom, not for evolution per se.

William Poteat and Wake Forest College enjoyed two years of relative peace. In 1925, the Baptist anti-evolutionists tried again to purge evolution by condemning Poteat and the college. The state convention that year was in Charlotte. On the day it began, writes historian Suzanne Linder, the anti-evolutionists

Creationism in North Carolina

had a clear mandate from their home associations to press the issue, and a clear majority to achieve it. But many Wake Forest alumni rushed to Charlotte to support Poteat, easily outnumbering the fundamentalists when the convention resumed the next day. In fact, the convention declared in a resolution that "we interpret the record in Genesis, not as myth, but as God's inspired revelation. We believe that it is literal and unassailable as to the fact of creation by God" (*Biblical Recorder* [Cary, N.C.], 25 October 1980:13). Yet it rejected any direct criticism of either Poteat or Wake Forest College, thereby signaling that it was permissible to teach evolution in the Baptist schools (Linder 1963:151–152).

How important was this event? According to Linder,

> This Charlotte Convention put Wake Forest in a new light. A Baptist college dared to acknowledge the teaching of evolution, and the alumni were willing to fight to protect this right. Faced with such strong alumni support, the fundamentalists revised their position. . . . Few of [the alumni] cared much about the theory of evolution. Some did not even understand it thoroughly, and certainly most of them were devout churchmen. But they remembered that Wake Forest had opened for them the wonderful world of the mind, and they did not intend to have that door closed in the faces of their sons. (Linder 1963:152–153)

While the Baptist anti-evolutionists faltered, their Presbyterian counterparts launched a second campaign. Governor Cameron Morrison, a Presbyterian, rejected two biology textbooks proposed for the public schools, saying, "I don't want my daughter or anybody's daughter to have to study a book that prints pictures of a monkey and a man on the same page" (Gatewood 1966:106). Subsequently the North Carolina Academy of Science passed a resolution defending Darwinism from Governor Morrison's views (Troyer 1987:47). The Presbyterian leaders then directed their attention to conditions at the University of North Carolina, where they assumed a proprietary interest: "So many Presbyterians occupied positions of influence

152

within the University during the nineteenth century that it was sometimes described as an agency of the Presbyterian church" (Gatewood 1966:104). In the legislature, another Presbyterian, David Scott Poole, introduced a bill to prohibit evolution in all tax-supported schools. At the legislative hearings, the central issue was the teaching of evolution at the University of North Carolina. Two distinguished members of the North Carolina Academy of Science, B. W. Wells and Zeno P. Metcalf, testified against the bill (Troyer 1987:48). The Poole bill passed in the Education Committee, but when it came to a vote on the floor of the state house of representatives on 19 February 1925, legislators who were alumni of UNC or Wake Forest voted solidly against it, defeating it decisively (Gatewood 1966:146). Sam J. Ervin, Jr., who would later become the U.S. senator who presided over the Watergate hearings, was a first-year state representative from Morganton at the time. He recalled the debate on the Poole bill as the most interesting debate of his entire career, and in a speech during that debate he suggested that "the passage of [the Poole bill] would do good in one respect. It would give joy to the monkeys in the jungle for the North Carolina legislature to absolve them from responsibility for the conduct of the human race in general and the North Carolina legislature in particular" (*News & Observer* [Raleigh, N.C.], 21 November 1981).

The Presbyterian Synod of North Carolina would not let the matter rest. It sustained the morale of the anti-evolutionists by passing seven anti-evolution resolutions at its annual meeting in October 1925 (*News & Observer*, 26 July 1964, sec. 3:2). Those forces regrouped in the spring of 1926 but quickly suffered three major setbacks. First, their organizing convention was an utter fiasco, featuring confusion and intolerance, giving the movement a farcical reputation. Then T. T. Martin came to the state to lead their campaign, but the North Carolina fundamentalists shunned him, throttling him as effectively as he had hoped to throttle evolution. Finally, they pinned their hopes on the Democratic primary of 5 June 1926, demanding that candidates take a stand on evolution. Most of those who opposed the teaching of evolution lost heavily. "The public generally interpreted the outcome of the primaries as a serious, if

Creationism in North Carolina

not disastrous, setback for the anti-evolutionists" (Gatewood 1965:289).

Poole was reelected, however, and he introduced another anti-evolution bill, but it was crushed in the house's Education Committee. "By the spring of 1927, the discussion of evolution had virtually ceased in North Carolina" (Gatewood 1966:229).

To clarify the factors that shaped that episode of anti-evolutionism, it should be noted that a parade of nationally known fundamentalist leaders marched through the state to inspire the faithful—T. T. Martin, Mordecai Ham, Cyclone McLendon, William Jennings Bryan, and Billy Sunday—but none could convert their considerable fame into tangible results. Instead, the state resolved this controversy according to its own parochial cares. Who respected William Louis Poteat; how organized were the fundamentalist Baptists from the mountain districts; what was the financial condition of the Baptist colleges; who had graduated from UNC or Wake Forest: these were the considerations that determined the votes at the Baptist conventions and in the state legislature.

Another important consideration was the Baptists' stance against evolution. Their fundamentalists devoted themselves almost exclusively to putting their own house in order, that is, to purging evolution from Baptist schools. But the civic principle that church and state should be separate was quite precious to them, so much so that they stayed away from the fights in the state legislature to control the public universities. This they left to the Presbyterians. Ironically, however, the anti-evolutionist Baptists influenced the rest of the state unintentionally. When they failed twice to rebuke evolution in their own schools, this was a sign that the forces opposing evolution were suspiciously weak (ibid.:178). If the Baptists could not purge evolution even from their own denomination, how could less conservative churches achieve anything?

Next, one should be cautious not to translate the victories over anti-evolutionism into endorsements of evolutionary thought. The Baptist State Conventions of 1922 and 1925 voted for church harmony and academic freedom, not for Darwinism per se. Likewise, in the legislature's votes against the two Poole bills, "The opponents of the legislation could rarely be classified

154

as either evolutionists or modernists. For the most part they were moderates, often theological fundamentalists, who preferred to have their faith compete in the free marketplace without the aid of legislative props rather than risk a violation of the principle of separation of church and state" (ibid.:231).

Evolution after the Second Poole Bill

The record on the place of evolution in North Carolina public school science courses after 1927 is mixed. The educational policies recommended by the State Department of Public Instruction (DPI) in Raleigh were consistently friendly to Darwinism, but Darwinism was largely purged from science textbooks because of out-of-state pressures.

The first major statement on science education subsequent to the Poole bills was DPI's science curriculum guide for 1930. At first glance, evolution seemed weak and religion strong in this document, as if anti-evolutionism had achieved its goals. Its section on geology includes the statement that "the remains of plants and animals found in formations have a historic interest" but avoids saying where that interest is likely to lead. The guide's biological concepts emphasize continuity and minimize change, for example, by stressing that life comes from life and that young living things resemble their parents. Also, this document erects an elaborate moral framework around its scientific topics. One of its spiritual goals is for the student to acquire "an attitude and desire for obedience to the law of love" (North Carolina State DPI 1930:415, 396).

These opinions were the DPI's roundabout way of saying that education is more a matter of building character than of acquiring knowledge. Given that philosophy, evolution was no more neglected than any other scientific topic, and science as a whole was no more subject to moral platitudes than were history, English, or physical education. When the 1930 science guide turned from its homilies to the nuts and bolts of what to teach, evolution reappeared as an important scientific topic. The DPI had rejected the theory that immorality could be traced to the idea of evolution.

The next science guide, for 1935, listed these statements

among its important scientific principles: "The earth is constantly changing," and "many types of life that existed in the past are now extinct." It supported them with these "scientific facts": "Each period of change lasted millions of years," and "all life has evolved from simple, pre-existing forms and has certain marks of similarity" (North Carolina State DPI 1935).

The 1941 guide was nothing less than an explicit, adamant, and detailed statement of evolutionary science. It began by quoting Herbert Spencer regarding the importance of science education, and John Dewey concerning the social consequences of scientific knowledge, after which it presented the concepts that "have *appreciably governed* the selection of subjects in elementary science programs" (North Carolina State DPI 1941:25 [original emphasis]). They include these:

- "The earth has developed as a result of the action of natural forces."
- "All life has evolved from simple forms."
- "Species have survived because by adaptations and adjustments they have become fitted to the conditions under which they live."
- "Through interdependence of species and struggle for existence, there is maintained a balance among the many forms of life (ibid.)."

Furthermore, these principles appeared in the same curriculum guide as organizing concepts for many individual units within the yearly plans for each grade. Unit 2 of fifth grade science addressed this information: "The changes in rocks, records in fossils, and extinct plant and animal life tell us that the earth is millions of years old. (About 300,000,000.) Extinct animals . . . tell us that life upon the earth has changed" (ibid.:85). For unit 5 of seventh grade science, students should learn that "there is a continual struggle for existence between plants, between animals and between animals and plants. . . . The healthiest and hardiest of the species usually survive" (ibid.:103).

Thus uniformitarianism, adaptation, variation, competition, natural selection, and so on were tightly woven into North Carolina public school science education, as conceived by the state education bureaucracy. And, instead of being isolated in high school biology, they appear in the fifth grade curriculum and in subsequent levels, which means that students were expected to recognize the general outlines of evolutionary science by the time they entered high school.

The curriculum guide of 1953 reaffirmed the 1941 principles without elaborating upon them (North Carolina State DPI 1953), but the next major document, the 1958 guide, named several items of Darwinian and Mendelian evolution that deserved special attention. Students should know "what is meant by natural selection, . . . variation, adaptation, survival of the fittest, artificial selection, [and] the theory of mutation." Another important topic was "What was early man like?" Recommended reference sources included T. Dobzhansky's *Evolution, Genetics, and Man*, W. Howells's *Back of History: The Story of Our Origins*, and E. A. Hooton's *Up from the Ape* (North Carolina State DPI 1958:47, 26).

These science curriculum guides indicate that North Carolina's top education officials felt that evolution deserved a major role in science education. Yet the guides were never controlling documents in the sense that the state's public schools had to follow them faithfully. Local school boards conformed voluntarily, or not at all, or voluntarily to some parts while ignoring others. A conservative school board could direct its science teachers to omit the sections on evolution, or to rush through them quickly. Field trips and science fairs might displace units on evolution, while units on other topics, such as cell theory or personal hygiene, could be expanded to take attention away from evolution.

Furthermore, if science textbooks were unfriendly to Darwinism, the policy concocted in Raleigh was subverted. After the Scopes trial, textbook publishers largely edited Darwinism out of their science texts because numerous local school boards around the nation feared the kind of controversy that Dayton, Tennessee, had experienced (Grabiner and Miller 1974). Sometimes the publishers concealed evolution beneath a blanket of euphemisms and ambiguities, while other times they dropped it entirely. Such were the textbooks available to the North Carolina public schools.

Creationism in North Carolina

It is easy to suppose that North Carolina treated evolution the way its neighbors did. After all, it was a southern state, a Protestant land, a poor commonwealth with a fragile system of public schools.

What, then, was North Carolina's experience in the classrooms? Based on my interviews with creationists whose high school educations in the state's public schools spanned a time from 1930 to 1976, I conclude that some North Carolina school boards suppressed the teaching of evolution, that some others may have presented it in a domineering way, but that many found middle-of-the-road compromises, balancing explicit teaching of evolution with religious tolerance for those who could not accept its assumptions (Toumey 1987:413–416).

If newspaper coverage is any reflection of the strength of an issue, then organized opposition to evolution was of little consequence in those years. The only mention of it in the clipping files of the *News & Observer* is a pair of short articles from 16 and 17 April 1965 describing an anti-evolutionary complaint by Rev. Ray Fuller of the Goldsboro Ministers Association, who formed a committee to investigate whether the teaching of evolution in the local schools was contrary to religious belief. Subsequent developments, if there were any, earned no mention in the newspaper. Incidents of anti-evolutionism were sporadic and isolated, so that it never achieved much importance as a statewide issue.

In one notable incident, however, evolution was too hot to handle in the public schools of Gaston County, west of Charlotte, in November 1971. George I. Moore III, a UNC graduate who was doing his practice teaching to earn his teacher's certificate, served as a substitute teacher in a Gaston County junior high school. On the tenth of November he found himself in the midst of a lesson, based directly on the words of a history textbook, about "how the belief in special tribal gods changed gradually to the belief in one God," that is, Judaism. Moore referred to this gradual change as the "evolution" of Hebraic religion, which ignited a freewheeling discussion, consisting largely of the seventh graders interrogating him as to his personal beliefs about evolution. He responded that he believed in Darwin's theory, that he was an agnostic, and that, in his view, some

158

parts of the Holy Bible "should not be taken literally" (357 F. Supp. 1037 [1973]).

That night the superintendent of schools heard from the angry parents of some of the seventh graders. Two days later, after a cursory investigation of the incident, Moore was discharged.

George Moore sued the Gaston County Board of Education to be reinstated. The federal district court judge who heard his case agreed with both of the plaintiff's arguments: that the principle of academic freedom entitled Moore to due process, and that the Board of Education had violated the First Amendment separation of church and state. Judge McMillan wrote:

> In view of the bloody history of tyranny and ignorance which had so frequently followed the union of Cross and crown, the Founding Fathers were obviously interested in freedom *from* religion of state origin or sanction, as much as freedom *of* religion of their own choice. . . . If a teacher has to answer searching, honest questions only in terms of the lowest common denominator of the professed beliefs of those parents who complain the loudest, this means that the state through the public schools is impressing the particular religious orthodoxy of those parents upon the religious and scientific education of the children by force of law. The prohibition against the establishment of religion must not be thus distorted and thwarted. (357 F.Supp. 1043 [1973]

Finally, a study by Elaine C. Grose revealed how students in North Carolina felt about evolution in 1976 (Grose and Simpson 1982). Grose tested a sample of 120 students in an introductory biology class at a large state university in North Carolina. For 80 percent of her sample, this was their first college-level science course, so that, in effect, her results reflect attitudes formed before these students entered college. She reported this distribution:

- 54 percent believed in evolution; this group included 9 percent of the sample who believed *strongly* in evolution.
- 24 percent expressed anti-evolutionary beliefs; here, 5 percent of the total sample was *strongly* against evolution.
- 22 percent were either neutral or doubtful about the credibility of evolution.

If events between 1922 and 1976 set the stage for North Carolina's creation-evolution controversy in the 1980s, then creationist sentiment was surely widespread, but it was not sufficiently powerful to banish evolution from science education or from public images of science. Creationist and evolutionary thought coexisted in a stalemate. Each had considerable support, but neither suppressed the other. Also, creationist sentiment took the form of a moral theory about evolution, namely, that Darwinism was directly responsible for immorality. Absent was the claim that the plenary authority of science substantiated the stories in Genesis.

Nine
Modern Creation-Evolution
Controversies in North Carolina

North Carolina's record on the creation-evolution issue contained too little drama and conflict to merit much coverage when the national rise of modern creationism came to the attention of North Carolina's newspapers in the early 1980s. There was much coverage of nationally known confrontations like the Arkansas trial of 1981, and some vague speculation that this issue would soon infect the legislature and the public schools, but there was scant mention of real creationists actually advocating creationism. The *Charlotte Observer* on 21 September 1980 documented public school policies in Lancaster, South Carolina, that included scientific creationism, but it cited no such cases for North Carolina. It could only report that a fundamentalist leader, Rev. Dan Carr, "said legislation to introduce the teaching of creationist theories [in North Carolina schools] is likely within the near future." Even so, Carr admitted that other issues like opposition to abortion and the Equal Rights Amendment were more important at the time. Two years later, after the creationist cause had received national attention while the Arkansas law was passed, challenged, tried, and struck down, there was still next to nothing newsworthy to report about creationism in North Carolina. The Greensboro *Daily News and Record* of 16 May 1982 quoted the complaints of Rev. Lamarr Mooneyham, chair of North Carolina Moral Majority, regarding the teaching of evolution in the public schools:

> The irony of the whole thing is that there was never
> a referendum or formal change in the educational

process. The process of change was itself evolutionary until, over a period of years, evolution became the exclusive explanation for the origin of man. Biblical creation was somehow removed . . . I am irritated as well as just insulted that we are not allowed to be part of the pluralism in public education. I pay taxes, too.

The same newspaper article balanced Mooneyham's comments with anticreationist statements by Rev. W. W. Finlator, a liberal Baptist preacher, but its most interesting information on the state's creation-evolution controversy came from Paul H. Taylor, director of science education for the Department of Public Instruction. Taylor thought that creationism had no place in the science curriculum: "Creationism is, in my opinion, a religious concept. It is not a science . . . The basis for evolution is science. If both are presented in the same context in a classroom in the pretense that they are both science, I think both religion and science are weakened or misunderstood." He said, however, that his office had no official policy on the teaching of creationism because "there is no overwhelming pressure we can detect in the state to force the issue."

One reflection of creationism's dormant condition in this state was the meager attention that the national leaders devoted to North Carolina. The Institute for Creation Research rarely mentioned North Carolina in its chronologies that record the lectures, debates, sermons, and seminars conducted by Henry Morris and Duane Gish. Between 1974 and 1984, it cited only five visits to North Carolina. Morris debated William Pollitzer, a physical anthropologist, at the University of North Carolina in Chapel Hill in September 1974; Morris and a colleague conducted a seminar in Raleigh in March 1980; Morris visited Reidsville, Greensboro, and Winston-Salem during a five-day visit in October 1983; Duane Gish spoke at a Christian high school in High Point in March 1984; and Morris went to Winston-Salem for four days in September 1984 (Morris and Gish 1976:5; Morris and Rohrer 1982:226; *Acts & Facts*, December 1983, June, November 1984). To these can be added a brief visit by Rev. Jerry Falwell in May 1981 specifically to criticize the exhibits on evolution at Durham's North Carolina Museum of Life and Science; a May 1984 appearance by A. E. Wilder-Smith, a European spokesperson for creationism, in the Research Triangle area; and an ICR-sponsored summer institute at Piedmont Bible College in Winston-Salem in August 1985 and 1986.

If indeed the national leaders neglected North Carolina, what does that say about the locals' tie-in to the national movement? Three occasions—from 1981, 1983, and 1984—shed light on this question.

Rev. Jerry Falwell came to Durham on 4 May 1981 to film some footage for his weekly television program, the "Old-Time Gospel Hour." He intended to broadcast his views on evolution and creationism in June, using exhibits on primates and paleontology at the North Carolina Museum of Life and Science as backdrops. With him came Lane Lester, the leading creation-scientist at Liberty Baptist College, who supplied technical details for Falwell's comments.

The museum permitted Falwell to film its exhibits, but its director, William M. Sudduth, pointed out to a local newspaper that he disagreed with Falwell (*Durham Sun*, 5 May 1981). When I spoke with him on 26 October 1981, he mentioned also that the museum frequently receives flyers protesting its evolutionary family tree mural that links humans with apes. Furthermore, he said that a local group had proposed a creationist display to accompany the museum's evolutionary exhibits, citing these problems with the current exhibit: (1) the fossil record does not necessarily support evolution; (2) radioactive dating methods do not provide good indicators of the age of the earth—since God could have made different levels of radioactivity, rates of decay could have been different in the past; and (3) God made human beings pure in spirit, meaning that they could not be related to lesser forms of life.

Sudduth also told me that these contacts with Falwell and the local creationists had not changed anything at the museum. Although he agreed to receive whatever literature they sent him, he told them he did not intend to alter the exhibits by including creationist information, especially their opinion that the age of the earth was ten thousand years. To illustrate his scorn, Sudduth told me, he threw away the local creationists' letters. The consequence of these contacts was that the museum's exhibits appeared briefly on Falwell's television program in June 1981, but the local creationists had no effect in altering anything at the museum in Durham, with or without Falwell's help.

Another creationist event was Duane Gish's talk at First Wesleyan Church in High Point on 30 March 1984. This church sponsored a broad array of ministries, which together constituted an impressive religious community. On the church grounds were a nursing home, a retirement home, a bookstore, an elementary school, a high school, and numerous minor facilities. When Gish spoke there, the high school was running a weekend youth assembly for 250 students, and Gish's presentation was part of the opening-night program held in the school's gymnasium on Friday evening. About 200 adults, presumably parents of the students, also attended. Almost everyone there was white.

First Wesleyan's youth pastor, Mark Welch, introduced Duane Gish, commenting that evolution is "a great lie that's being taught to people," and hinting that his success in inviting Gish had been a miracle of God, since he was able to reach Gish directly even though he is such a busy man that he is seldom in his office when people call. Three high school boys sitting behind me in the audience snickered at this miraculous coincidence.

Holding a Bible in his hand, Gish spoke first of his boyhood in White City, Kansas, a small farming town. After telling how he and his brother once stopped a train by greasing the tracks, he said that he accepted Jesus Christ while in the tenth grade, whereupon his language, his friends, and his pastimes changed. The Bible is inerrant, he said, and he had found no scientific fact that contradicted it. Gish alleged that most scientists are unbelievers who begin with wrong assumptions, for example, mechanistic or naturalistic ideas. Labeling evolution "a twentieth-century mythology," Gish told the audience that "evolution is chance. Don't let anybody kid you. Ultimately, it all comes down to chance." Although the greater part of his talk was a diatribe against evolution, Gish offered a few ideas

164

in support of creationism. He explained that fossils could be understood in biblical terms since the "behemoth" of Job 40:15 was a dinosaur that had become extinct after Noah's Flood.

"Don't be ashamed of the Bible," said Duane Gish to the high school students. "Some teacher will tell you that you have evolved. Don't believe it." After speaking for about forty-five minutes, he finished by urging the young people to commit themselves to Jesus Christ.

Gish had presented himself and his message smoothly, and the audience at First Wesleyan had seen why he is creationism's top orator. Masterfully he wove together scientific statements with biblical belief, thereby giving the gathering reason to conclude that science corroborates conservative Christian values. And yet there was something not quite right about the listeners' response to Duane Gish that night. The evening's program had begun with taped hard-rock music playing in the gym as the students arrived. These young people were dressed more or less as one might expect from public high school students, a few quite neatly but many much more casually. Some wore ripped T-shirts and had unkempt long hair. Before Gish spoke, a young Christian singer named Amy Gunden entertained the gathering. There was nothing gospel about her music, however; it was simply rock music with superficial Christian lyrics. Between songs, Amy Gunden spoke vaguely about the insecure feelings of young people like herself, obliquely adding that "people let you down, but if you put your security in Jesus, He won't let you down." Gunden sang again after Gish completed his talk.

While Gish spoke, the irreverent boys snickered, gossiped, and yawned. Only four or five people there, besides Gish himself, had Bibles with them. In many ways, this occasion was almost as much mainstream secular youth culture as it was religious: the rock music, which conservative Christians abhor; the sloppy clothing, which they equate with degeneracy; the watered-down Christian witness of the singer; the few Bibles, a peculiar omission for a creationist venue. Except for Gish's speech and the introductory comments by the youth pastor, this evening had the ambience that fundamentalists expect to find at a wishy-washy middle-of-the-road church. That being so, it seems Duane Gish delivered his views in a doctrinal vacuum, for the high school students at First Wesleyan lacked the harsh animosity toward evolution and secular values that fuels creationism. No one there disagreed with his creationism, but few showed much concern about it.

A third occasion, revealing a much stronger bond between a national creationist leader and his audience, was Henry Morris's appearance at Community Baptist Church in Reidsville, in October 1983. During his five-day visit to the Greensboro area, he preached in a Friday night service at Community Baptist, then conducted four one-hour sessions on creationism at the same church on Saturday. This appearance, incidentally, had a special sentimental meaning for Morris, for the pastor at Community Baptist, Rev. Jim Dotson, had helped him twenty years earlier in Blacksburg, Virginia, when Morris left a moderate Baptist church and founded an Independent Baptist church. Dotson's Community Baptist Church in Reidsville is also an Independent Baptist congregation.

The Friday night venue at Community Baptist drew about a hundred people, representing a wide spectrum from teenage to elderly. Most came as families. All were white. The service began with two hymns by the church's excellent choir, followed by a solo hymn. Then a missionary, Norm Niemeyer, spoke for a half-hour about his adventures in Asia and elsewhere, exciting the audience with a tale in which each frustration he faced was the product of an anti-Christian conspiracy, whether of Communists, Hindus, or Muslims. When Niemeyer finished, another singer sang a solo hymn, after which Morris spoke. His topic was "Evolution in Turmoil," the title of one of his books. Unfortunately, however, the hour was late, so Morris had to rush through his talk in twenty minutes, speaking mostly in broad generalizations.

The next day, Saturday, about sixty people attended each of the four sessions of Morris's presentation on creationism. Many people were present at all four, though some came only for the two morning talks, and others only for the afternoon periods. There were approximately as many women as men in his audiences, and in fact many people came as couples. Each unit began with a prayer, and hymns were sung between units.

Henry Morris chose 2 Peter, chapter 3, as his biblical text for the day ("All things continue as they were from the beginning of creation"), using its verses to introduce creationist information, and using his creationist findings to illuminate biblical exegesis. There was no distinction made between biblical belief and scientific creationism, because both were part of Morris's message that day. He said that the first law of thermodynamics meant God has finished the creation, while the second law meant that Jesus is conserving God's creation, and "in the New Testament, [entropy] means confusion and shame."

Much of Morris's presentation consisted of quoting such famous evolutionists as Julian Huxley, Theodosius Dobzhansky, and Stephen Jay Gould, to hang them by their own words. From the audience came frequent indignant murmurings against the evolutionists' intolerance and inconsistencies, at least as exposed by Morris, and righteous enjoyment of the evolutionists' discomfort at Henry Morris's hands. Still, he softened his criticisms of the atheists, theistic evolutionists, or evangelical Christians who dissented from his views by reminding the gathering that these people were reputable scientists.

Between sessions I heard people from the audience express profound respect for Morris. Said one man, "I wouldn't want to have to debate anyone *that* good!" This was slightly ironic, for Morris's personal demeanor was modest, even shy, when in one-on-one conversations. Even when talking from the podium, he was quite soft-spoken and mild mannered. But these people appreciated him as a teacher, not as a spellbinding orator, and admired him accordingly.

The last of the four units that day, a question-and-answer session, revealed the reactions of the local conservative Christians to Henry Morris's work. A few of their questions were focused on technical topics, and others on biblical meanings, but most combined the two spheres. One person asked when the dinosaurs were born or, rather, created. Morris answered that they were created in the six-day creation like all other animals, and that the "behemoth" and "leviathan" mentioned in Job 40 and 41 were probably a brontosaurus and a marine dinosaur, respectively. Another person inquired whether Job actually saw dinosaurs, whereupon someone else asked what would happen if live dinosaurs were discovered today. A series of questions were asked about Noah's ark: Why did Noah take two animals from some species, but seven from others? What happened to the clean beasts versus the unclean ones? Did Noah and his family eat fish while on the ark? How did they keep their food fresh? Likewise, two Garden of Eden inquiries: If all was harmonious at first, when did animals start eating each other? When the lion lies down with the lamb, will this be like the Garden of Eden? Matters of creationist chronology drew two questions: What about the problem that some stars are billions of light-years away if the universe is only a few thousand years old? Can we gauge the age of the universe from the depth of dust on the moon or the age of comets?

Others wanted to know Morris's opinion about how Lot's wife actually turned to a pillar of salt, and what it meant when Gen. 10:25 spoke of the division of the earth in the days of Peleg. Finally, questions in the realm of morals and ethics posed these problems: Is genetic engineering creating new life, and, if so, should it? Is cloning life-forms ethical? What are the moral implications of evolution outside of science, say, in education and politics? How can evolutionary ideas in social science be countered in topics like cave dewellers or New World migrations?

I cannot do justice to all of Morris's replies except to say that he answered the technical questions by referring to Bible stories, and he responded to the Bible topics by presenting technical information. The former kind of answer is the one that bothers the anticreationists, for whom creationism is making science conform to biblical belief, as when Morris accounts for dinosaurs in terms of "behemoth" and "leviathan." But the other aspect should not be overlooked: sometimes Bible stories are forced to fit into scientific terms, as Morris's answer to the question about Lot's wife shows:

> What happened to Lot's wife? It says she looked back and she became a pillar of salt. Well, of course, you have to accept what it says. There have been various explanations of it. One is that as she sort of hung back she didn't really want to leave, 'cause she liked her home in Sodom and she didn't want to leave it. So

she kind of lagged behind and therefore she wasn't with Lot and the daughters. And so in the eruptions, land went up like a smoke or a furnace, it says, and there is evidence that there have been volcanic eruptions, and there are great salt deposits around there. But maybe she was just sort of buried in a layer of volcanic ash or something, and maybe that she was buried in a shower of salt, although that's not too likely, I don't think. More likely, maybe she was buried in this ash, and over the years, just like a buried body may become petrified, fossilized, by chemical replacement, atom by atom, she gradually became a pillar of salt. That's what Harry Rimmer suggested years ago in his book called *Lot's Wife and the Science of Physics*. Other than that, other than just a pure miracle, I don't know. God didn't want her to lag behind, obviously.

Whatever the moral meaning of the story of Lot's wife, the chemical replacement theory complicated it, for it logically implied that one's understanding of God's message was contingent upon secular knowledge. Admittedly, Morris was winging it with answers like these. His chemical replacement scenario is hardly a central feature of creation-science. Even so, it might be objected that he can't have it both ways. He can judge natural science by biblical standards, or he can justify the Bible according to the standards of natural science, but to do both together is circular reasoning. Yet what seems circular to an observer appears seamless to a creationist. As far as that day's listeners were concerned, all things fit together perfectly after Morris explained them, and the story of Lot's wife became more real when it acquired scientific sanctification as "chemical replacement" was invoked. More than anything else, the gift these Christians received from Henry Morris was the message that one can be perfectly comfortable with Bible stories, taken literally, and with scientific data, however dense, at the same time, because the two constitute one truth. Actually, most of Morris's listeners that day were as naive in scientific matters as they were expert in their Bible study; the sheepish ways they posed their technical questions showed this. But that did not matter so much after a day with Henry Morris, for they drew from him the confidence that science isn't all that intimidating if you read your Bible and use it to resolve scientific issues.

Contrast Henry Morris's day in Reidsville with the equivocal consequences of Falwell's visit and Gish's. The latter two failed to produce any noticeable changes in the creation-evolution controversy, but Morris's had the modest effect of intensifying local creationist belief. This was not a qualitative transformation, as if evolutionists were converted to creationism at Community Baptist Church; rather, it was a case of preaching to the converted, which was just as important in its own way. While the visits of the famous experts may have been both flattering and exciting, they were not doing much more than fine-tuning the sentiments of the local creationists.

In addition, the comments and questions at Morris's visit to Reidsville indicate that, by the early 1980s at the latest, the appeal to scientific sanctification had been well cemented into creationism in North Carolina. Creationism was no longer restricted to the theory that learning evolution leads to sin. It had also acquired the central message of Morris's ministry, namely, the expectation that scientific authority would authenticate the stories of Genesis. Creationism was then understood to be *scientific* creationism.

Public School Controversies

In North Carolina, as in the nation as a whole, creationists felt it was urgent to get *scientific* creationism into the public school science curriculum, for both symbolic and instrumental reasons. To have it included in science education would be prima facie evidence that creationism was scientific, and that it was entitled to all the secular respectability that comes with science; also, its inclusion would be a good way to inculcate the next generation with the lessons of creationism. Schoolchildren need not be taught Bible stories per se in the public schools, but if they were taught that creationism is scientific, then the lessons they learned in Sunday school and in the home presumably would be that much more convincing.

The first recorded instance of modern creationist agitation in North Carolina was Rudene Kennedy's presentation to the state textbook commission on 8 October 1974 asking that creationist materials be used by the public schools (as opposed to simply banning the teaching of evolution, which was the creationist strategy earlier in the century). Kennedy was the Creation-Science Research Center's representative in North Carolina. "The response of the committee," wrote Nell Segraves of CSRC, "was excellent" (Segraves 1977:16). The textbook commission did not reject any evolutionary materials, however, and it certainly did not adopt any creationist books.

Three years later, an ad hoc group called Christians for Academic Freedom approached the Charlotte-Mecklenburg schools regarding a balanced treatment of creationism and evolution. Speaking for the group, a schoolteacher, Elizabeth Machen, complained that children were being "indoctrinated" in evolution. A lawyer representing the creationists, James Atkins, argued that "the theory of evolution is on the verge of collapse," and he said that several school systems were currently adopting "equal time" policies (*Charlotte Observer*, 20 April 1977; cf. Segraves 1977:17). The school board never endorsed any such policy, but in that year, 1977, it faced a controversy over evolutionary statements in an earth science textbook titled *The World We Live In*. After parents and a preacher voiced complaints at a school board meeting, the board provisionally withdrew the book and sought guidance from the state Board of Education, which decided there was a religious conflict in the content of the text. Charles Vizzini of the Charlotte-Mecklenburg schools told me in a telephone interview on 22 November 1983 that the Charlotte-Mecklenburg board then discontinued that book.

In the years thereafter, there was occasional mention of creationist sentiments and creationist activities around North Carolina. Students for Origins Research, in California, mentioned a creation-evolution debate at Mount Olive College, a Free Will Baptist school in Wayne County, and a seminar on creationism at Southeastern Baptist Theological Seminary in Wake Forest (*Origins Research*, Spring 1980:8). When local creationists brought Henry Morris to Raleigh for a weekend seminar

Creationism in North Carolina

in March 1980, two local papers reported the event and mentioned the views of those who sponsored Morris (*News & Observer* [Raleigh, N.C.], 22 March 1980; *Leader* [Research Triangle Park, N.C.], 17 April 1980).

When proposals to mandate equal time for creationism appeared in dozens of state legislatures in 1981, one was introduced in the North Carolina state senate. Jack Cavanagh was a first-year Republican from Forsyth County who had been elected in the Reagan landslide of 1980; he was also active in the Christian Action League, a fundamentalist lobbying organization loosely affiliated with the State Baptist Convention. His bill, "An Act to Require Balanced Treatment of Creation-Science and Evolution-Science in the Public Schools," was a carbon copy of Arkansas's Act 590 in its definitions of creation-science and evolution-science, as well as the policies it required for "lectures, textbooks, library materials, or educational programs that deal in any way with the subject of the origin of man, life, the earth, or the universe" (North Carolina State Senate DRS11578–LE, 1981). Among the findings of fact it proposed were that "public schools generally censor creation-science and evidence contrary to evolution," that "evolution-science is contrary to the religious convictions or moral values or philosophical beliefs of many students and parents," and that suppression of creation-science "undermines [students' and parents'] religious convictions and moral or philosophical values, compels their unconscionable professions of belief, and hinders religious training and moral training by parents." Then, developing the First Amendment establishment of religion argument against evolution, it proposed to correct evolution's illicit educational monopoly by giving creationism equal time.

Senator Cavanagh's bill died in committee without attracting a single cosponsor, either Republican or Democrat. I met with him at the General Assembly on 9 June 1982 to ask him about his initiative, and he said he was not aware that his proposal was identical to Arkansas's law, nor had he been in contact with any national organizations, whether Moral Majority, Institute for Creation Research, or Creation-Science Research Center. Cavanagh related his experience:

I got additional information for my research from people around here who had heard the idea of [my] bill, and they had written to [ICR] at one time, so they were bringing in some materials to help me. Newspaper articles, magazines, but it wasn't any formal organization at all. You know, it's strange. After you introduce a bill like that, you'd expect that those that are interested in doing some changing, such as the Moral Majority, such as the [creationist] group out in California, would contact you and say, "We can help" or "Here's what you'll be up against," but I got absolutely no call, no letter, from any of those organizations. . . . I am not in contact with the Moral Majority, nor with Pat Robertson, or PTL, or any of those groups, not because I wouldn't be delighted to talk with them, but they haven't gotten in contact with me. I'm not being funded by any of them. I've been accused of that, but it's not true.

Cavanagh told me in our 1982 interview that he tried to contact someone locally who he had heard was interested in the creation-evolution issue, but that he was never able to reach that person, whose name he had since forgotten. Thus, except for the informal contacts he mentioned, he received no support at all. After that bill flopped, he said he tried another approach:

The second bill I put on the calendar was a bill to create a research study commission to investigate, in the state of North Carolina, in our schools, what is being taught as far as the origin of the species is concerned. . . . I went to the North Carolina Textbook Depository warehouse with three other people, and we looked through all the anthropology and biology books we could find there, and not one dealt with creationism as a method of describing how we came into being.

That bill also went nowhere. Senator Cavanagh speculated that he should convince some conservative Democrats to cosponsor his creationist legislation if he tried it again in the future, and that he should try to steer it to a sympathetic committee, but he was defeated in the 1982 election. No creationist bill has arisen since then in the state legislature.

Although Cavanagh focused his attention on evolution and creationism in those bills, he told me in our 1982 interview that he was also very much concerned about the moral theory that evolution is a part of Secular Humanism:

> There is a religion in this country called Secular Humanism. What I mean by that is that it has been described as a religion, determined a religion, by two Supreme Court decisions. And evolution is a basic tenet of the Secular Humanist religion, in that if they don't have evolution, then its whole premise falls apart, because they believe that no deity can save us, we must save ourselves, that there is no God, that man is the beginning and end of all things, man himself. So when you take away evolution, then you destroy it because they have to account for something coming from nothing. . . . I've noticed that men are losing a little bit of their self-esteem, their manhood, if you will, their desire to achieve, to grow, to develop, some of that stuff inside a man that's supposed to make him feel—it's almost like the Indians, where if you take away their right to hunt, they become less of an Indian. Well, I'm beginning to see that in man, and it's frightening. It is truly frightening. But where I really see it is in the church. The church is losing its backbone. It's losing its commitment. And they are frightened from taking a stand on what they supposedly are to believe. . . . You're heading for a collision, a collision between Secular Humanism and Christianity. If man needs to be brought back up to a position of good, and the only way he can do that, in the Christian religion, is to be redeemed, then he needs a redeemer. And that's Jesus. Which is hogwash to the Secular Humanists, who believe that man has never fallen from grace and is getting better and better. So,

what the Secular Humanist religion is saying by virtue of its sticking hard and fast with evolution is that there is no Jesus, no need for a redeemer. So you've got this tremendous clash between faiths.

As with Cavanagh's complaint, the usual pattern of anti-evolutionary activity was not to focus exclusively on the idea of evolution, but rather to include evolution as one item within a broad range of objectionable topics or materials that, collectively, constituted Secular Humanism. For example, Moral Majority of North Carolina published a booklet titled "Textbook Reviews" in April 1981. After scanning numerous textbooks, novels, and other materials, Moral Majority cited passages that were especially colorful or objectionable, asking its readers to "consider by comparison how far our system has digressed from a time when reading, writing, and arithmetic were guaranteed to our students and twelve years of education bore the same equivalent." Following sections on sex education and health that denounced information about adolescent sexuality, the review's brief section on biology focused on abortion, for instance, statistics of spontaneous abortions and a description of vacuum curettage, with only an oblique reference to evolution—in the context of abortion—plus a complaint about an idea from sociobiology: "Advocates of this school of thought have found it easy to accommodate not only abortion, but phenomena such as war, exploitation of one's fellow man, colonialism, and racism. The fact that some animals control their numbers by killing their young should not compel us to do the same." The only other direct reference to evolution and creationism was a list of ten "basic beliefs of humanism," one of which stated that humanism "denies the biblical account of creation" (North Carolina Moral Majority 1981).

A source from the opposite end of the political spectrum also suggested that anti-evolutionary grievances constituted only a minor feature of conservative Christian concerns, nested within a much broader understanding of U.S. culture and morality. The North Carolina branch of People for the American Way (PAW), a group established to oppose organizations like Moral Majority, studied censorship by surveying teachers and administrators in North Carolina public schools during 1983 (North Carolina Project 1983). From a sample of 467 replies, PAW reported that 26 percent of the state's public schools "have experienced one or more challenges to books or instructional materials since 1980." Saying there had been at least 243 censorship attempts, the PAW document described the most common style as "an effort to ban a particular book or magazine from the classroom or library shelf." Furthermore, "the prevailing reasons for attempting to ban these books were objections to profane language, discussion of human sexuality and discussion of atypical lifestyles." The report stated explicitly that "the teaching of evolution was a significant source of censorship activity," but it counted only 10 such cases within its total of 243, of which "at least three . . . resulted in limitation of the teaching of evolution, and in one additional case the child of the complaining parent was removed from the class." No doubt there could have been more cases than PAW discovered, but four incidents of censoring evolution over three years was less than catastrophic. A similar measure of anti-evolutionism can be extrapolated from the report's statement that 13 percent of the challenges, or 31 out of 243, addressed an entire subject, as opposed to a single textbook or magazine. Of those 31 cases, "almost all . . . were discussions of human reproduction in health classes or other forms of sex education. The remainder were objections to the teaching of evolution." Thus, the document by North Carolina People for the American Way yielded the same conclusion as the publication by North Carolina Moral Majority: despite the fuss about evolution and creationism in newspaper headlines and television news, these issues played a minor role, relative to problems of sexuality, in the concerns of the state's fundamentalists.

When public schools receive complaints about a curriculum or a textbook, the most important factor for resolving the complaint is the curriculum guide, that is, the statement of teaching goals and methods prepared by the state Department of Public Instruction and approved by the state Board of Education. Curriculum guides are revised approximately every five or six years, after which publishers are invited to submit textbooks that conform to the guides. The Textbook Commission then asks teachers in the appropriate subjects to judge the books both for quality and for conformity to the guide. It is not unusual for a thousand or more teachers to evaluate a particular textbook (*News & Observer*, 11 March 1984, 14 December 1986). After receiving the readers' reviews, the Textbook Commission recommends several books in each field to the Board of Education, which usually approves at least three in each field. The local school boards have the freedom to use whichever books they want, but the state subsidizes only the ones approved by the state Board of Education, which means that the local boards almost always use the state-approved books. This process generates three important results. First, the state's public school teachers, because they have been deeply involved in the selection process, are disposed to using and defending the state-approved books. Second, there are alternative textbooks for most subjects; if a given text faces minor objections, it is possible to substitute a different book while still conforming to the standards of the curriculum guide. Third, it is extremely expensive for a local school board to adopt a text that the state has not endorsed.

Next in the process for dealing with curriculum complaints are local school board procedures. For most of North Carolina's school districts, of which there are about 140, only a parent of a public school student can make a formal complaint to the school board. Preachers cannot represent parents, and parents who have withdrawn their children from the public schools to enroll them in fundamentalist Christian academies cannot instigate challenges. This removes some of the schools' severest critics from the decision-making process. When a legitimate challenge does arise, the school board can allot several months for discussion and debate. Instead of resolving a complaint in a climate of urgency, when opinions are polarized, the board can wait while numerous diverse points of view emerge, a process that encourages consensus decisions and centrist policies. Finally, the local school officials can isolate a parent's complaint by making an exception for a single student; they can assign an alternative textbook for the daughter or son of an unhappy parent or, in extreme cases, can excuse the child from the objectionable units of a course. Meanwhile, the rest of the students continue to learn the subject within the framework of the state's curriculum guide.

Procedures like these had been adopted by almost all of the state's school districts, but many teachers and librarians were unaware of them when a wave of complaints from fundamentalists hit the public schools in the early 1980s (North Carolina Project 1983:5). Within a few years, however, the procedures were well known and used often. Said A. Craig Phillips, superintendent of public instruction, in November 1983, "we went through a time when a lot of fundamentalists were looking at school materials. Most school systems did a very strong job of setting up a way in which people with a concern could express it. I feel good about what we've done. . . . My impression is that there's been a minimal amount of unrest. Censorship efforts, on either side—left and right—have been very limited" (*News & Observer*, 7 November 1983).

A study by Mary Ann Weathers, assistant superintendent for curriculum and instruction in the Moore County school system, confirms Phillips's comments. In 1986, Weathers sent a survey to the state's 140 school superintendents and received 129 responses. She concluded that "every school system has a procedure on file for handling challenges [although, says Weathers, not all personnel knew this]. . . . The procedures proved effective and rarely were materials modified or removed or techniques changed or deleted." In three-fourths of the complaints, "there was no change to the curriculum because of the challenge" (Weathers 1986). Similarly, a 1985 report by the national office of People for the American Way mentioned curriculum complaints by conservative Christians in Buncombe, Carteret, and Mecklenburg counties, but the complaints produced no changes (People for the American Way 1985).

Outside the public schools, there was a loosely organized opposition to creationism. Two scientific organizations reacted against Senator Cavanagh's proposed bill by adopting resolutions. The North Carolina Science Teachers Association published a position paper in September 1981 that conceded that evolution should be taught as a theory, "not a fact," and that it should not be presented dogmatically. It also said, however, that, "in general, creationism is a religious concept. Religion is

178

based on one's belief or faith, not on scientific evidence. Evolution is a scientific theory based on scientific data accumulated over many years and organized, by logic and reason, into a unifying idea. The theory of evolution is, as all theories are, tentative in that it cannot produce a conclusive answer" (North Carolina Science Teachers 1981).

The North Carolina Academy of Science in 1982 also passed an anticreationist resolution. Unlike the science teachers' statement, the academy's position conceded almost nothing about evolution being just a theory. On the contrary, it said that "debates about evolutionary mechanisms . . . [are] a normal part of how science works. . . . Debate about evolutionary mechanisms in no way undermines scientists' confidence in the reality of evolution." Furthermore, it asserted, "no scientific hypothesis suggested as an alternative to evolution has succeeded in explaining relevant natural phenomena. Moreover, insights provided by evolutionary principles have been the basis for progress in the biological and biomedical sciences which has benefited mankind in many ways." In its summary, it declared that "the North Carolina Academy of Science strongly opposes the mandated inclusion of creationist views of origins in public school science classes. Furthermore, the Academy is strongly opposed to any mandated exclusion of the principles of evolution from public school instruction" (North Carolina Academy 1982).

The state's major newspapers also opposed creationism, and especially its proposals for equal time in science classes. Most vociferous were the *Daily News and Record* of Greensboro, and the *News & Observer* of Raleigh. On 5 and 13 March 1981, the Greensboro paper denounced the creationist philosophy represented in a California court case involving the Creation-Science Research Center; it also applauded the decision in the Arkansas trial on 8 January 1982. The *News & Observer* expressed its disdain of creationism on the eve of the Arkansas trial, on the centenary of Darwin's death, on the occasion of an anti-evolutionary decision in Texas, when the Randolph County school system canceled an educational play about evolution, at the publication of a theistic evolutionist book, and when a group of Nobel laureates joined the Louisiana case in an amicus brief

Creationism in North Carolina

(23 November 1981, 24 April 1982, 25 January, 29 March, 23 July 1984, 24 August 1986).

Whether these anticreationist proclamations influenced the creation-evolution controversy in North Carolina is hard to say. The state's fundamentalists were most unlikely to heed the advice of the *News & Observer*, the paper they loved to hate, and the resolutions of the two scientific associations were almost as obscure as Senator Cavanagh's proposals. Still, creationist achievements in public education were rare. One such victory was won in March 1984, when school officials in Randolph and Mecklenburg counties decided that a play about evolution for elementary school children was too controversial to stage (*News & Observer*, 28 March 1984; *Charlotte Observer*, 29 March 1984). *Dandelion*, a humorous piece that included fish becoming amphibians and primates becoming bipedal, was often performed in North Carolina schools by an educational acting troupe from Charlotte. It had been booked by ingenuous planners in the Randolph schools, but school officials heard a story on National Public Radio in February 1984 saying it was controversial because of its evolutionary content, whereupon they arranged to replace it with a play about Alaska and Hawaii. Interestingly enough, the fear of controversy arose among the school officials without any external prodding by creationist parents or conservative preachers. The officials were not so much advocating creationism as fleeing rumors of notoriety. The school superintendent, George R. Fleetwood, explained that his staff canceled the play because "they were alerted to the fact that it had caused concern in other cities." Another administrator, Loraye H. Hughes, commented, "I didn't want to present something that maybe some of the parents would have some hesitation about. Why do that when you can get a play that has no controversial overtones and it is as good a quality? Really, I was looking for a historical play to begin with, not a science play." Similarly, Barbara Koesjan of the Charlotte-Mecklenburg schools decided not to have *Dandelion* shown, explaining that "we serve such a diverse group of people, and I see no reason to raise that flag [evolution] with them. Maybe public people are running a little bit scared on some things" (*Charlotte Observer*, 29 March 1984).

180

The Orange County Textbook Dispute of 1984

A textbook controversy that arose in Orange County in 1984 shows the local flavor of North Carolina creationism. This county, in the east-central piedmont, is sharply divided into two social spheres, namely, a rural conservative zone to the north and west, and a cosmopolitan liberal community in the southeast. The latter is Chapel Hill, home of the University of North Carolina. This is one of the most politically liberal communities in the country; it voted for Walter Mondale in 1984 in about the same landslide proportions by which the rest of the nation voted for Ronald Reagan, that is, about 60 percent. But in the rest of Orange County, to the north and west of Chapel Hill, most of the rural precincts came in for Reagan (*Chapel Hill* [N.C.] *Newspaper*, 7 November 1984). Here is a landscape of dairy farms, tobacco fields, little towns clinging to minor textile mills, and old Baptist churches at nearly every intersection where secondary roads cross.

In 1982 a new conservative church was formed on the outskirts of Hillsborough, the county seat. This was Abundant Life Church, an independent charismatic congregation. Its pastor was Rev. David Smith, personable and articulate, and its membership included many middle-class families. In a telephone interview on 2 December 1986, a liberal critic, Rev. Richard Hildebrandt of the local Presbyterian church, described the members of Abundant Life as "traditionalist. . . . They find very simple answers to the questions around us."

According to Smith, a group of parents came to him in the spring of 1984 with complaints about the textbooks their children were reading in the local public school system (which is independent of the Chapel Hill public schools). He and those parents had three meetings at his church between then and October, by which time about a hundred people had come together with common complaints in this matter (*News & Observer* 19 October 1984).

Smith told me in a telephone interview on 2 December 1986 that, as the group and its concerns grew, they consumed much of his time; "it became so pre-occupying that I turned it over to a committee." The committee called itself Concerned

Citizens for Better Public Education. Most of its constituents were members of Abundant Life.

When this committee came to the attention of the local newspapers in the fall of 1984, Smith said that it was reviewing textbooks from the local public schools for "antireligious" and "anti-moral" content. "The ultimate goal," he said, "is to see that texts that are blatantly teaching antibiblical, anti-moral, anti-religious positions—to have them replaced—that's the ultimate goal." Concerned Citizens for Better Public Education would report its objections to the school board "within the next few weeks," he said (*News & Observer*, 19 October 1986).

The committee turned to Ann Frazier, the state's leading fundamentalist critic of secular textbooks, for help in finding evidence of Secular Humanism in the books. This was a notable move, because Frazier was an electrifying personality. Her voice was sweeter than Sourwood Mountain honey, yet from her mouth came startling statements about humanism, atheism, communism, and innumerable other evil isms. When I heard her speak about six months after those Orange County events, she alleged that someone conspires to keep children illiterate so they can be collectivized, and that the New Age movement, which she said plans to control all human thought, is promoting abortion, suicide, and euthanasia so that the world's population can be reduced to two billion by A.D. 2000. Those who shared Ann Frazier's views were often galvanized to action, while those who differed with her were typically appalled by what she said.

Excitement grew in September as June J. Haas, a conservative school board member elected in May, announced a search for Secular Humanism and humanists in the public schools. In a letter to the editor in Hillsborough's weekly newspaper, she said, "I believe that humanism is being taught in our system and have strong suspicions that we have active humanists working within our system" (*News of Orange County*, 19 September 1984). Also at about that time, said Hildebrandt, a liberal opponent of Haas, the fundamentalist parents' committee circulated a letter warning that "come this fall, many parents will unwittingly send their defenseless children back into the 'Temples' of the public classroom to be 'brain-washed' and pros-

elytized by purveyors of this 'new faith' (secular humanism)"
(*Durham Morning Herald,* 20 October 1984).

The Concerned Citizens for Better Public Education never
identified any single teacher, principal, or administrator as the
source of the Secular Humanism in the schools. Indeed, they
claimed it was rampant, but no one in particular was responsible
for it. Said Haas in a telephone interview with me on 2
December 1986, "The textbooks bother me more than the
teachers."

The hunt for humanism raged in the letters to the editor
pages of the local weekly. According to one, "Many alert parents
know that Secular Humanism has become the established religion of the U.S. public school system. They believe the Humanist Manifestos provide the key to unlock the code languages
of values clarification, progressive education, sexuality curriculums, situation ethics, and other thought processes that helped
eliminate prayer, moral training, and the teachings of basics
from the public schools" (*News of Orange County,* 26 September
1984).

Haas's letter to the editor of 19 September 1984 gave a
detailed definition of Secular Humanism: the denial of the deity
of God, of biblical inspiration, of the divinity of Christ, of the
existence of the soul, of life after death, of salvation, of heaven,
of damnation, of hell, of creation according to the Bible, of
absolutes, of a sense of right and wrong, and of distinctive roles
for males and females. Furthermore, continued Haas, Secular
Humanism advocates sexual freedom, including premarital sex,
homosexuality, lesbianism, and incest, not to mention abortion,
euthanasia, suicide, "equal distribution of America's wealth,"
control of the environment and of energy, plus removal of
"American patriotism and the free enterprise system." Lastly,
this thing that lurks in the public schools advocates disarmament
and "one-world socialist government."

Complaints about evolution arose in the wake of the alarms
about Secular Humanism. The conservative Christians' most
salient objection to evolution was that, as Rev. David Smith
told me in our interview, "the theory of evolution is being taught
as though it was proven with scientific fact." Smith gave me an
example by citing a passage from *Understanding Psychology,* a

textbook published by Random House. In that passage, the ability of infants to grasp objects was alluded to as a product of evolution, that is, from apes in the past who were able to grasp branches. This bothered Smith because, he said, "It is preassuming evolution as a fact." This same problem also occurred, he said, in a biology textbook and in some geography books: "Evolution is alluded to as if it were a fact." This assumption was particularly troubling to him because, he said, "the only reference to creationism was put in the context of a legend, a myth—to us that was very offensive."

As a remedy, Smith suggested to a newspaper that "the biblical account should be given equal weight with evolution in textbooks" (*News & Observer*, 19 October 1984). Haas echoed Smith's sentiments: "Who's to say that evolution is a more popular theory? Who can say that? I think creation should have equal time" (*Durham Morning Herald*, 17 October 1984).

In the comments of most of the conservative critics, the issue of evolution and creation was minor, relative to other complaints. Smith described it as "a major complaint" in our interview, but Cheryl C. Atwater, another leader of Concerned Citizens for Better Public Education, told me in a telephone interview on 3 December 1986 that "evolution was not the main factor." Ray Hooker, another of the textbook critics, insisted that "these parents may prefer that the subject of evolution versus creationism be given an intellectually honest treatment in the schools, but this issue is not the battle cry" (*Durham Morning Herald*, 1 November 1984).

If any topic was more salient than evolution, it was sexuality, especially when connected with abortion. In our 1986 interview Smith said this about the topic of abortion: "A section in a health book handled [abortion] in a positive context. There was no mention of the psychological repercussions. No apology. It was not handled fairly"; elsewhere he complained that "abortion is being talked about very freely. It is handled as an acceptable alternative to a full-term pregnancy" (*News & Observer*, 19 October 1984). In the same vein, Haas told me in 1986 that "students have been taken to the Health Department to get birth control," and that some textbooks have "pretty deep sex ideas, about how to stimulate your partner." Ray Hooker echoed this last complaint by writing that "The 7th grade biology textbook described in detail the techniques of sexual foreplay for arousing a woman for sexual intercourse and then comparing social aspects of dancing to those of sexual intercourse among teen-agers" (*Durham Morning Herald*, 1 November 1984).

Ira Trollinger, the assistant superintendent for instruction in the Orange County Schools, reacted to the conservative Christians' criticisms by defending the process of textbook selection, pointing out that it was both careful and thorough (*News & Observer*, 19 October 1984). This might seem as if he was dodging the controversies about the content of the textbooks, but his way of protecting the schools was to defuse divisiveness and bitterness. For this he had a recent precedent. About a year before this textbook controversy, the evangelical Christian businessmen's group known as the Gideons wanted to distribute free Bibles to students in the Orange County public schools. No doubt this suggestion could have led to a classic controversy of religion in the public schools, setting conservative Christians in the community against the school officials, particularly if the latter were contemptuous of or insensitive to the Gideons. The school officials were firm in declining to let the Gideons distribute Bibles but were also polite about it. The issue died without arousing further attention.

When I interviewed Trollinger in Hillsborough on 2 January 1984, about ten months before the textbook controversy became a public issue, he said he felt that evolution is an essential part of science education, but as a theory, not as a fact. And, while he was adamant in denouncing creationism's intrusion into the scientific sphere, he was also eloquent in endorsing it as a religious truth. He is a Christian, he said, and believes in a Divine Creator; also, he stated, many others in the state's science teachers' association are Christians too.

Trollinger described several ways to accommodate parents' concerns about evolution. Parents may ask to exempt their children from evolution classes (or sex ed or other troublesome topics), and the schools ordinarily comply. Trollinger added that, when he receives such requests, he urges parents to let the children learn about evolution *without having to believe in it*, thereby easing the conflict between the parents' values and the contents of a course or textbook.

What all this means is that the school system's assistant superintendent in the 1984 textbook controversy was sensitive and temperate. Trollinger knew how to oppose the more extreme demands of the conservative critics, but also how to respond to parents' more reasonable concerns.

One more party joined the controversy. Hildebrandt, the liberal Presbyterian pastor, formed a group called Parents for Public Schools to counteract the fundamentalists' committee, Concerned Citizens for Better Public Education. At that time, Hildebrandt said in our 1986 interview, "it appeared that a majority [on the school board] would support the desires of the Abundant Life Church."

Thus, by late October there were four principal parties, namely, an angry group of religious conservatives, a worried group of liberals, an elected school board that was probably leaning toward the conservatives, and a centrist administration. When it seemed as if all elements were in place to begin a long and difficult struggle for control of Orange County's public schools, an odd thing happened. The conservatives' committee intended to present its criticisms as an official complaint at a formal meeting of the school board, and so to demand certain changes in the textbooks. It was discovered, however, that the school board could only entertain complaints from parents who had children in the Orange County public schools. All other parties were, in effect, outsiders, e..luded from initiating changes. This meant that the conservatives' committee, as such, had no standing before the school board. Furthermore, the leaders of the conservative committee lacked the appropriate standing. Some were not parents; some were parents whose children had since graduated from the Orange County schools; some were parents with children in private religious schools. None had the credentials to bring formal complaints before the school board, let alone the authority to force changes.

Were these people the victims of an unreasonable technicality? Perhaps they felt so at the time; however, there is nothing particularly contentious about the principle that, in local issues of public education, the parents of the children in the

186

public schools have the primary rights, not to be superseded by the interests of other parties. Even if the Orange County school board was a little too severe about applying it, couldn't the conservatives have presented at least a few parents with the proper standing who were willing to press the complaints about Secular Humanism, evolution, and sex education?

Once that procedural problem arose, the conservative attack receded. The only development that could possibly be construed as a change of policy was the introduction of a moment of silence in the classrooms. Smith and Haas reported an intangible accomplishment, which was that they "were able to arouse the consciousness of other parents, and . . . got some parents talking about it," Smith told me. Still, Haas was not satisfied with those results, and Smith admitted in our interview that "it's difficult to get textbooks removed."

Since 1984, Abundant Life Church has started its own school, thereby directing more fundamentalist parents away from public school controversies. Also, Hildebrandt's liberal group had a friendly candidate elected to the school board, giving them a temporary majority. Hildebrandt told me in December 1986 that he sometimes hears that Abundant Life Church will reorganize its criticisms of the local public schools, but that he had not heard anything since late 1985.

In understanding this case, three aspects seem especially important. First, the business of evolution and creationism arose within the wake of more general and more powerful concerns about Secular Humanism. As an independent issue, creationism had little force of its own. Second, the Orange County public school system had a durable set of commonsense policies that discouraged drastic institutional changes. The flexible procedures that Trollinger described gave dissatisfied parents some reasonable options within the school system, while the school board's parents-only rule for curriculum complaints protected the system from outside parties. This rule worked against the fundamentalist group in this case, but, in the long run, it probably works equally against all external groups. Yet it is nicely balanced by the internal options that let individual parents, including fundamentalists, keep their children in this public school system while protecting them from a few objectionable features of the curriculum.

Finally, the educational alternatives developed by fundamentalist groups, especially home education and church schools, had the ironic effect of making things easier for the public schools by steering the most angry parents away. Many teachers regret that conservative Christian parents take their children out of the public schools, both because these schools lose some good students and because some church schools are substandard, at least as outsiders describe them. These circumstances are regrettable, but so are the furious accusations that public schools are cesspits of violence, promiscuity, and drug abuse. It is one thing for uninformed critics to hurl these charges at the public schools, but when angry parents within the public system say these things, then the system is torn by internal conflicts setting parent against parent, and parent against teacher.

Dénouement

By 1985, the creationist challenge to science education in North Carolina's public schools had almost evaporated. This can be seen in the comments of three people deeply concerned about the role of evolution in the science curriculum. Ira Trollinger, the assistant superintendent in the Orange County schools, had been president of the North Carolina Science Teachers Association in 1983. When I spoke with him early in 1984, he said that he had noticed very little statewide pressure from creationists; speaking as an officer of the science teachers' group, he told me that the organization was comfortable with the status of evolution at that time. Most biology teachers, said Trollinger, treat evolution as a basic biological concept, though they ordinarily represent it as a theory. (Here he was recapitulating the NCSTA's 1981 statement.) To the best of his knowledge, he continued, there were no conservative Christian groups pressuring the science teachers on the creation-evolution issue, and in fact there was little mention of it among those teachers.

Another interested party was Ray Flagg, a vice-president of Carolina Biological Supply, the manufacturer and publisher of materials for science education, whom I interviewed by telephone on 24 July 1985. Flagg had been president of the North Carolina Academy of Science in 1983–84. He helped draft the academy's 1982 position paper, and he debated a preacher on evolution and creationism that year. Also, he criticized creationism in the Burlington newspaper at about that time. For a few weeks afterward, he received several nasty telephone calls objecting to his evolutionist views, as well as some harassing calls at home in the middle of the night, with the caller hanging up when he answered. Flagg said, however, that the calls ceased long ago. His only other direct contact with creationism consisted of rare complaints by fundamentalist Christian schools regarding Carolina Biological Supply's books and films, which mention evolution prominently.

The third person, William Spooner, of the Division of Science in the North Carolina Department of Public Instruction, agreed in our telephone interview on 29 July 1985 that creationism had peaked. He said there were many complaints around 1981 and 1982, often involving threats of litigation, but, he added, "I haven't encountered it much in the last couple of years [1984 and 1985]." Spooner was confident that students whose teachers followed the state's curriculum guide for science were exposed to evolution, which, he said, was implied throughout the guide. He explained that local school boards beset by creationists could protect themselves by referring complaints to the state Department of Public Instruction, which could be represented by the state attorney general, since the school boards were merely following state guidelines when they featured evolution in their science courses. At any rate, said Spooner, the objections he noticed had shifted from the specific issue of creation and evolution to more general concerns about humanism, which were very vague and very global.

The comments of Trollinger, Flagg, and Spooner are impressionistic, but some tangible facts corroborate them. One, from February 1985, concerns biology textbooks. The national office of People for the American Way evaluated eighteen high school biology texts for their respective approaches to evolution. Although half were disappointing, in varying degrees, three were considered especially bad because they totally avoided the word "evolution," while three others were praised for their forthright presentations of evolutionary ideas. None of the "bad" books had been approved by the North Carolina State Board of Education, but two of the three "good" ones had. In the opinion of Manley W. Midgett, science supervisor for Wake County schools, "Textbooks used in the local schools address evolution and . . . all of the major books being considered for use next year [1986] addressed the subject" (*News & Observer*, 22 February 1985).

The 1986 study by Mary Ann Weathers of challenges to curriculum materials attested that "most [1985–86] challenges were in the areas of sex education, values clarification, and library materials." There was only one anti-evolution case (Weathers 1986).

The last corroborating item I will mention is the 1985 *Standard Course of Study*, a comprehensive public school curriculum for grades K–12 prepared by the Department of Public Instruction that emphasizes critical reasoning in the state's semiofficial philosophy of education. Conservative Christians, led by Ann Frazier, bitterly attacked the *Standard Course of Study*, accusing it of being, for a start, humanistic, anti-Christian, antifamily, and anti-God. In its section on the science curriculum, the *Standard Course of Study* tiptoed around the creation-evolution controversy by avoiding explicit mention of the E word, but it expressed its policy indirectly by including, among its learning objectives, the concept that "living things are a product of their environment" and by listing such topics as heredity, adaptation, uniformitarianism, tectonic forces, geological time, change through time, and fossils in its outlines for eighth grade and high school science (North Carolina State DPI 1985). One of the conservative Christian critics, Marcell Souther, focused on its implied endorsement of evolution, which, she alleged, taught students that "they are animals." She continued, "We ought not be surprised when students 'act' like animals and call their public school a 'zoo.' The message has gotten through to them, and they are behaving in a manner faithful to the concepts of those who conceived their education" (*News & Observer*, 4 April 1985).

The State Board of Education was scheduled to endorse the *Standard Course of Study* in Spring 1985, at about the same time that the state legislature was considering the Basic Education Program, a grand plan to reorganize the financing of North Carolina's public schools. Although the two policies required separate decisions, with the Board of Education passing judgment on the *Standard Course of Study* and the legislature enacting the financial program, they were informally linked in the sense that the legislature could reject the Basic Education Program if it was greatly dissatisfied with the board's philosophy of education. It was the hope of the conservative Christians that the legislature's hearings on the Basic Education Program could be used to criticize the *Standard Course of Study*, and to alter it by mandating curriculum changes before passing laws for financing public school education. Legislators listened politely to opinions like Souther's, then, after making a few minor changes, heartily endorsed the Basic Education Program without altering the *Standard Course of Study*.

Ten
Inertia and Centrifuge:
The Paradox of Sectarian Support

The peculiar thing about creationism's recent history in North Carolina is that its conservative Christian friends were so thoroughly unsuccessful in pressing the argument that scientific authority supports the creationist view. Even though creationist knowledge and belief were solidly institutionalized in fundamentalist churches, schools, and other associations, nevertheless the state Academy of Science, the science teachers' association, the local school boards, and the state's education officials all firmly rejected the creationist arguments.

To untangle this puzzle, we can revisit the two existential themes of modern creationism in the United States and ask how they crisscrossed in North Carolina. The constituency of conservative Christians was fertile soil for the moral theory that evolution was somehow responsible for a host of evil developments. The state Moral Majority's "Textbook Review" denounced evolution in the context of condemning abortion; Abundant Life Church in Hillsborough tied evolution to abortion and sex education; Senator Jack Cavanagh placed evolution under the giant dark shadow of Secular Humanism; and Mary Ann Weathers's report on school censorship, like those of People for the American Way, portrayed opposition to evolution as a small part of a much larger moral complaint. Even Duane Gish's preaching against evolution at First Wesleyan Church in High Point was part of an evening dedicated to the moral development of Christian youth.

That being so, the main question was whether the considerable local sympathy for the moral theory about evolution

The Paradox of Sectarian Support

could be converted into creationism's stance on scientific authority: that is, could scientific sanctification be derived from the religious complaint? Would scientific creationism achieve secular respectability by being included in the science curriculum of the public schools, on the grounds that Darwinism was immoral?

At the very least, it is plain that the moral theory could not be readily translated into scientific authority. And yet these two existential themes were not exactly independent of each other, either. They were intimately entwined within the religious culture of North Carolina, but, ironically, the way they were entwined was that sectarian support *diminished* scientific credibility. The more adamantly the conservative Christian organizations insisted that evolutionary thinking was immoral, the more the rest of the North Carolina population distrusted the creationist agenda for science education.

To understand that paradox, we have to consider the state's sectarian politics. "In North Carolina, only the sparrows outnumber the Baptists," says a local adage. During the early 1980s, the State Baptist Convention, a unit of the Southern Baptist Convention, had more than three thousand member churches, embracing more than 1.3 million adherents, who together represented 22.8 percent of the state's population, or 42.3 percent of all people with religious affiliations in North Carolina (Quinn 1982:21). (The State Baptist Convention, incidentally, did not include most black Baptists, Independent Baptists, Free Will Baptists, or Primitive Baptists, who together constituted several hundred thousand additional Baptists.) Every major issue with moral or religious implications, including the creation-evolution controversy, had to come to terms with the fact of North Carolina's Baptist plurality. So great were the Baptist proportions that if the state's Southern Baptists spoke with one voice, then the State Baptist Convention could have easily pushed its views through the state legislature, whether on abortion, school prayer, creationism, or anything else.

But Baptist politics did not work like that, for there were several critical mechanisms that prevented factions of Baptists from converting their moral sentiments into enforceable policies. One mechanism was the autonomy of local congregations,

193

by which Southern Baptists meant that no one has any business telling anyone else how to run their congregation: not one's fellow Baptists, nor the State Baptist Convention, and certainly not the national Southern Baptist Convention. Indeed, the state and national organizations could pass whatever resolutions they liked, but they could not require any particular congregation to implement them. Thus there was widespread sentiment in recent years that women should never be preachers, and that converts should always be baptized by total immersion, and that using tobacco is sinful. Southern Baptists recommended these positions by passing resolutions at their national or state meetings. Some Baptist congregations took them to heart, but others reserved the right to ordain women, or to admit converts without requiring total immersion, or to condone the local tobacco economy, each time justifying themselves by referring to the principle of autonomy.

Another factor was decision by consensus. A group as large as the State Baptist Convention inevitably embraced many different kinds of people. While fundamentalists were the most vocal of Baptists, there were also many moderates and a few liberals. A plurality of members was moderately conservative: they were conservative in the sense that they respected traditional morality, yet moderate in that they rejected the shrill voices and divisive tactics of other fundamentalists. In addition, there were other institutions within the State Baptist Convention that exerted a moderating influence on Baptist affairs, notably Wake Forest University in Winston-Salem and the *Biblical Recorder*, the state convention's biweekly newspaper. That periodical was almost entirely devoid of debate on evolution and creationism during 1981, 1982, and 1983, the glory years of modern creationism—no editorials, no articles, only three letters to the editor (of which one embraced theistic evolution).

The editor of the *Biblical Recorder* was Rev. R. G. Puckett, a leader of the moderate conservatives, and also Jerry Falwell's number-one nemesis in the state. Puckett often denounced Falwell's motives and methods in the editorial pages of the newspaper. I asked Puckett in an interview on 8 November 1983 in Cary, North Carolina, why there was so little interest in creationism and evolution in the *Biblical Recorder*. Puckett said that it truly reflected the disinterest of his paper's subscribers. There had been no court cases in the state during this time, and no notable controversy in any particular school. (Puckett is not in any sense an evolutionist. "I'm a creationist," he said. "I believe the book of Genesis is God's revelation of the initiation of the universe by a Supreme Being." He vehemently separated himself from the scientific creationists, however, by arguing that God's act of creation must be accepted on faith, and not as a scientific theory.)

Fundamentalists felt that the *Biblical Recorder* and Wake Forest University were bastions of liberalism. These institutions were moderate by most standards, but they nevertheless served as a foil to the fundamentalists by framing Christian values in terms of moderation and consensus. Thus, because of its configuration of internal diversity, the State Baptist Convention of North Carolina often made middle-of-the-road decisions that were most agreeable to moderates and moderate conservatives, thereby frustrating fundamentalists.

The third feature that prevented conservative opinions from becoming public policy was the principle of church-state separation. The formative years of the nation's Baptist history included numerous instances of persecution by the established churches. In response, Baptists acquired strong reasons to appreciate the First Amendment freedom of religion principles, to the point that they enshrined church-state separation as a Baptist sacrament. Their traditional response to a moral issue was to be a Christian model by exhibiting moral behavior, but also to discourage all denominations, including their own, from lobbying the government directly.

To summarize: the mechanisms of local autonomy, consensus decision making, and church-state separation combined to neutralize the largest religious organization in North Carolina regarding civic implications for moral issues. Instead of taking decisive action that might change public policy, the State Baptist Convention maintained a moderately conservative inertia that discouraged political initiatives. No doubt this was enormously frustrating to fundamentalist Christians, who knew that many Baptists sympathized with their moral concerns, yet who could not translate that sympathy into political power or secular scientific

Creationism in North Carolina

authority. Some fundamentalists had faith that the State Baptist Convention could be saved from its own moderation, but many others abandoned it to form their own Independent Baptist congregations or to join other autonomous conservative churches, such as Bible chapels. The same was true for fundamentalist Christians in other Protestant denominations, so that the theological topography of North Carolina featured a kind of sectarian centrifuge, with unhappy conservatives casting themselves away from the central religious bodies and complaining on the periphery about moral issues, or rather the refusal of their less conservative coreligionists to tackle moral issues. Just as church-state separation was a Baptist sacrament, so sectarian schism was a fundamentalist ritual. Ironically, then, those who were most intensely concerned about common sinfulness dissipated their strength in the bitterness of their self-imposed isolation. While they considered the large denominations like the State Baptist Convention too wishy-washy to confront urgent issues, the large denominations thought the fundamentalists were too intolerant to be trusted on urgent issues. Thus the fundamentalists were trapped between the frustrating inertia of the moderately conservative churches and the political isolation of their centrifugal schisms.

Operating within those considerations were three conservative Christian organizations that the creationist movement could have considered its natural friends: the Christian Action League, Churches for Life and Liberty, and Moral Majority of North Carolina (Oberschall 1984; Oberschall & Howell 1982). Christian Action League (CAL) was (and still is) a lobbying organization located three blocks from the state capitol in Raleigh. Although its membership embraced more than a dozen denominations, it was most closely affiliated with the State Baptist Convention, its biggest contributor. Its executive director, Rev. Coy Privette, was a Southern Baptist preacher who was also a past president of the State Baptist Convention. More recently, CAL established strong connections with ultraconservative Republicans; for example, Rev. Privette aided the 1984 reelection campaign of Senator Jesse Helms by lending it the CAL mailing list (for which the State Baptist Convention reprimanded Privette). Privette was also a member of the state

196

house of representatives from 1985 to 1992. CAL focused its lobbying efforts on three traditional Baptist issues, namely, drinking, gambling, and pornography. On each it was often successful, easily uniting fundamentalist Baptists with moderate conservatives to support state laws and local policies that suppressed those three vices. For the more sweeping moral issues like abortion or creationism, the Christian Action League was unable to do much. Privette explained to me, at an interview in the Raleigh office of CAL on 8 November 1983, that he had had no direct contact with the creationist movement, either locally or nationally, when creationism was a burning issue in the early 1980s. For information on this issue, he said, "our basic source is the Bible." Indeed, Privette felt that public schools should teach scientific creationism alongside evolution, and he monitored the sorry status of Jack Cavanagh's creationist bills as they languished in uncaring legislative committees in 1981. But other issues like drunk driving and a proposed state lottery took up so much of his time that creationism became a relatively minor issue to the Christian Action League.

Churches for Life and Liberty (CLL) offered a different kind of support for creationism. This group represented a network of parent-teacher associations from many of the fundamentalist Christian academies in North Carolina. Its constituent schools taught creationism as natural science or as religious doctrine or both, so there was no ambiguity as to CLL's creationist sympathies. CLL turned its attention almost entirely to the mission of protecting Christian schools from state interference, however, as opposed to projecting Christian values into the public schools. This mission crystallized early in 1979, when the state legislature was considering a bill to hold private schools to certain standards of accreditation, for instance, by requiring that teachers be college graduates. The state's Christian academies interpreted this as an attempt by the state to take them over; led by CLL, they mobilized hundreds of people on short notice to demonstrate in front of the legislative building, whereupon the legislators lost their nerve and killed the bill. Since that time, CLL has been ever-vigilant in opposing state standards for teacher accreditation, day-care conditions, and policies for spanking students. The government has no business regulating private schools in any of these matters, according to Churches for Life and Liberty. For creationism, this meant that science education in the fundamentalist schools was safely shielded from the evolutionary implications of the curriculum suggested by the Department of Public Instruction. But CLL's defensive posture—pulling the wagons into a circle, so to speak—also meant that parents and teachers were isolated from public school issues, including the effort to introduce scientific creationism therein. What happened in the public schools was relatively unimportant to them as long as the integrity of the Christian academies was preserved. And when they took an interest in their local public schools, they were nakedly exposed to the charge that parents who abandoned public education had no business controlling its content.

The same was true of Christian parents involved in home education. A decision by the North Carolina Supreme Court in May 1985 sanctioned the practice of teaching one's children at home, whereupon thousands of conservative Christian families began to do so. By removing their children from all schools, even the Christian academies, they could present creationism and other fundamentalist topics whatever way they liked, but they drove themselves farther to the periphery of civic life, isolating themselves not only from the public schools but also from the Christian academies.

The third organization was Moral Majority of North Carolina. Its chair, Rev. Lamarr Mooneyham, was occasionally quoted for his support of creationism by the state's newspapers and television stations when the issue was in the news in 1981 and 1982. This seemed impressive at the time, as if Moral Majority was going to put its political muscle into the creation-evolution controversy, but that never happened, probably because Moral Majority of North Carolina had no political muscle. Like the national Moral Majority organization, North Carolina's branch was a mailing-list operation, not a network of activists. Its members participated mostly by sending donations to the leader and relying on him to take care of moral issues. Except for an occasional letter to to the editor from Mooneyham in the local papers, the only noticeable activity by Moral Majority of North Carolina during the crucial years of the creation-evolution controversy was the textbook critique of April 1981, which mentioned evolution in passing. For the next two and a half years, Mooneyham's organization existed more in the fears of Jerry Falwell's enemies than in the political realities of North Carolina. Then, about a year before the 1984 election, Mooneyham left the Independent Baptist church he had founded in Durham and moved to Charlotte to organize a voter registration drive among conservative Christians on behalf of Senator Jesse Helms. (It is likely that his efforts made a critical difference when Helms defeated his opponent, Jim Hunt, by about seventy thousand votes.) Yet the cause of creationism never benefited much from the activities of Moral Majority of North Carolina.

A choice of two strategies confronted the creationist movement: it could concentrate on convincing political authorities to include scientific creationism on equal grounds with evolution in the science curriculum; or, it could devote itself to restating its moral messages to its natural friends, that is, the conservative Christian churches, schools, and families that already accepted biblical creationism as inerrant truth. Preaching to the heathens about how things should be versus preaching to the converted about how things are. Choosing one did not necessarily preclude the other, but it was worth setting priorities. For the creationist activists of North Carolina, the choice was obvious. It could be taken for granted that conservative Christian institutions accepted the truth of God's creation and understood the evil of evolution, so the next logical thing was to project this truth into the rest of society, particularly the public schools. Thus the moral theory was basic, and from it the stance on scientific authority would somehow be established.

Taking creationism into the secular world required some political skill. How do you present your message to legislators, teachers, school boards, and public education bureaucrats so that they will agree to grant it a measure of secular credibility? For this, the creationists needed the help of organizations like the Christian Action League and the Moral Majority, which supposedly knew how to get public policies changed. Unfortunately, however, the Christian Action League had its hands full with a modest repertoire of issues it could realistically affect, which were drinking, gambling, and pornography. It certainly sympathized with the creationist cause, but turning public school science education upside down, as creationism required, was beyond its abilities. Moral Majority seemed like the next best group for getting things done, and in fact both Jerry Falwell and Lamarr Mooneyham often embraced creationism's proposals for equal time in the public school science curriculum. The disappointing truth was that Moral Majority of North Carolina was a paper tiger, however, especially in public school curriculum controversies. Although the media loved to quote Jerry Falwell and his North Carolina surrogate when they said outrageous things about volatile controversies, including creationism, the tangible changes these men desired were much more difficult to achieve than a mention in the morning headlines or on the evening TV news. Again, the natural friends of the creationist movement lacked the ability to force this issue in secular society.

To illustrate the difference between the difficulty of changing the external world and the satisfaction of consolidating one's influence among friends, compare the status of Bible study in the Reidsville public schools with the matrix of Christian values at Friendship Christian School in Raleigh. Reidsville, a small town in the north-central piedmont, north of Greensboro, had a policy of including Bible study classes in its public schools since 1922. A local coalition of conservative churches, the Committee for the Teaching of Bible, appointed the teachers and paid their salaries. The committee claimed that the classes complied with constitutional guidelines because their content was strictly historical and literary, but in fact their daily proceedings included prayers and hymns. In October 1982, William F. Horsley, a local attorney with a daughter in fifth grade in the Reidsville public schools, objected to the school superintendent that the Bible study classes represented an unconstitutional intrusion of religion into the public schools. The attorney for the school board affirmed Horsley's opinion. The school board then invited the Bible study committee to suggest alternative arrangements; instead, the committee organized a show of political strength, with petitions and a large turnout at the December 1982 meeting of the school board, where one person spoke

The Paradox of Sectarian Support

against the Bible classes, and about two dozen spoke in support of them. Regardless, the school board had the constitutional issues on its mind, and it voted three to two to end the classes (*News & Observer* [Raleigh, N.C.], 19 December 1982).

A month later, in January 1983, the sponsors of the Bible classes acquired a bus to be used as a mobile classroom near the public schools, but not on school property (*News & Observer*, 21 January 1983). This, presumably, would comply with constitutional law. Even so, the advocates of the Bible classes had lost a serious battle of symbols. Whereas previously their religious curriculum had occupied a prominent place in the local public schools, now it had been removed to the outer edges of legal propriety and spatial proximity.

The views of Rev. Jim Dotson, pastor of Community Baptist Church in Reidsville, help illuminate the dilemma of the creationists. Dotson was a longtime friend of Henry Morris, and he had also been on the Committee for the Teaching of Bible. I visited him at Community Baptist on 14 October 1983. After the Bible study dispute had been resolved, he said, he was "not at all" satisfied with the end result:

> I think we should have fought that battle in the courts if necessary. The school board did not. . . . All the school board had to do was say, we're going to continue teaching this until a federal judge says we can't. But they took the easy way out. . . . I think we should have stood up and fought the thing as far as necessary for the sake of the nation, not just Reidsville. . . . We might have lost . . . but at least we would have done what we could have. We settled for a compromise that was satisfactory to *some* people, but *I'm* not satisfied with it.

Similarly, Dotson's hopes for creationism in the local public schools were disappointed. Although this issue had no major lobbying effort comparable to the case of the Bible classes, Dotson told me in our interview that "we started our Christian school here, and part of the reason . . . is because we wanted our children exposed to biblical creation. And they would not get that in the public schools. So that's a part of the reason for having Community Baptist schools here."

In contrast, the officers of Friendship Christian School in Raleigh were well past the point of being frustrated by conditions in the public schools. They and their church, Friendship Baptist, had erected their own moral environment, grades K–12, which freed them from worrying about Bible study, evolution, or prayers in the outside world. Charles Stanley, principal of this school of 437 students, described the cultural conditions there in an interview at the school on 6 October 1983. Their value system was based on biblical inerrancy and salvation theology, he told me, which were transmitted and reinforced through numerous media, including these: Bible classes at all levels; textbooks written by a fundamentalist ministry in Pensacola, Florida; Pro-Teens, a Christian youth ministry for high school students; Awanas, a Christian youth group paralleling Boy Scouts and Girl Scouts; Sunday school; and Sunday sermons. Said Stanley, "There are no secular subjects in our schools. For the Christian, *every* subject is a sacred subject, because God is the author of all things, and all knowledge."

Thus, health was a Christian subject, teaching that God made our bodies; math was a Christian subject, concerned with "precision in numbers in the universe"; science was a Christian subject in which God governs the universe; government was a Christian subject in which one learned of humanity's failure to create a "great society"; and history was a Christian subject, teaching that Christ is coming back to establish perfect peace.

Meanwhile, the humanist philosophy of the public schools, according to Stanley, had no absolute values, standards, or truths. Public school textbooks dealt with death education, values clarification, and a "do-your-own-thing kind of approach," he said. But at Friendship Christian "we have a traditional approach, that is, we have a teacher, a class, we have rules, and guidelines, and regulations that the students go by. We don't have drugs, alcohol, tobacco, smoking, or rebellion in general."

According to the creationist content of Friendship Christian's curriculum, one learned in biology that God made the world in six literal twenty-four-hour days; in physics, that thermodynamic entropy began when sin entered the world; in social studies, that wars and rumors of war arise because humankind has rejected God's truth. As to evolution, Stanley argued that it "would tear down the family. The homosexual gay rights movement is very evolutionary. The women's movement is very evolutionary. The civil rights movement is very evolutionary. All these things have their roots in evolution."

In Stanley's understanding, creationism was integrated into all morals, homilies, judgments, and behavioral expectations at Friendship Christian School. By contrast, the outside world, including the local public school system, was thought to be a confused maze of ignorance, evolutionism, and depravity not worth reforming.

In October 1983, Dotson had recent memories of trying unsuccessfully to project creationism and Bible study into the local public schools, while Stanley was entirely devoted to the task of nurturing a separatist counterculture of fundamentalist schools and churches wherein creationism and Bible study were well appreciated. (Dotson had also established a fundamentalist school at his church and presumably would achieve what Stanley had accomplished, but that fall Dotson's policies were far less separatist than Stanley's.) Their respective circumstances, in Reidsville and in Raleigh, recall the choice of strategies that faced the creationists of North Carolina: to project the message into the secular culture around them, or to consolidate creationist belief within fundamentalist institutions.

For the local activists of the creationist study group that I describe in the next chapter, the greater part of their effort after 1983 was directed toward sympathetic conservative Christian groups. In their year's end letter summarizing their activities of 1983, they reported that "the objective for this year was to concentrate mainly on carrying the message of creationism to Christians. The Lord has opened many doors for us to move forward in this area." A month later, in January 1984, the organization's leader reminded the members that they would concentrate their efforts on apologetics and Christian education, as opposed to public meetings or debates, partly because the secular occasions drew "antagonistic individuals." The leader then cited his favorite example of ministering to a friendly Christian audience: when he and another member of the group had been invited to speak about creationism at Southeastern Baptist Theological Seminary, they were apprehensive because they felt unprepared, but when the Lord helped them make a successful presentation, they saw this as a sign that they were doing the work of the Lord. Through the rest of 1984 and into early 1985, as he and others in the study group mentioned their recent activities, they reported many visits to conservative churches in Dunn, Creedmoor, Raleigh, Apex, New Bern, and Roanoke Rapids, which together included Baptist, Pentecostal, and Presbyterian congregations. They recounted lectures to Christian youth organizations such as a Fellowship of Christian Athletes, an Assembly of God college student club, and a local chapter of Awanas. They made a handful of presentations to nonsectarian audiences, including my anthropology class at the University of North Carolina at Chapel Hill, but these were rare.

The question of proselytizing in the secular world versus consolidating in the conservative Christian culture arose at the February 1985 meeting of the local study group, when their main business was a report about a creationist association in a midwestern state, by a man who had been one of its officers. Mostly he talked about organizational matters, but he concluded with a lengthy discourse on the problem of religious issues. Whereas the scientists from the association tried to stick to scientific and educational matters, even when speaking at churches, he said, the preachers frequently raised contentious theological topics. One such topic that often arose was predestination, according to the speaker, while another was age at baptism, with Lutherans advocating infant baptism and Baptists insisting on the adult form. Since the creationist association included Pentecostal, Lutheran, and Baptist preachers, and the venues of their scientific experts included more than a hundred churches during a recent year, the problem of religious disputation was chronic. They never quite solved it, the man said; at best they kept it under control. One of the North Carolina creationists replied that if they ignored religion entirely, it would not be spreading the gospel.

Those last two comments captured the problem perfectly. When the creationist movement concerned itself with contentious religious topics, its message alienated people who were

The Paradox of Sectarian Support

not conservative Christians; but if it masked its spiritual meaning for the sake of secular credibility, then it betrayed its own soul. In the case of the North Carolina creationists, the scientific image had been a little more prominent than the religious facet until sometime in 1983, probably because the scientists among them had a sense of professional self-respect, and because the nonscientists deferred to the scientists' authority. Thus they avoided divisive religious disputes. Nevertheless, this strategy failed to convince secular authorities to include creationism in the public school science courses. As their frustration mounted on the secular front, these creationists withdrew into the network of their own sectarian friends, who always made them feel welcome. Yet a dangerous trap awaited them there, for the main concern of the fundamentalist churches was not scientific credibility. Rather, it was the threat of moral degeneracy in modern times, which presumably included Darwinism. Thus in the sectarian venues, the creationism message was a religious message delivered to a religious audience. Regardless of how these speakers felt about scientific credibility, the conservative congregations listened mostly for the moral content of creationism. They wanted to hear that their biblical standards were correct, and that other values were wrong. Thermodynamics were not nearly as important to them as denouncing abortion, nor was geochronology as urgent as opposing gay rights, nor could the paradigm shifts of Thomas Kuhn compete with the question of prayer in the public schools. While they no doubt appreciated the supporting role of scientific symbols that helped give creationism a respectable face, the last thing in the world they wanted to hear was that creationism could be made religiously neutral to make it fit neatly into the secular framework of public school education.

And so the creationist cause in North Carolina was enveloped in a deeply sectarian culture, with the paradox that the moderates and moderate conservatives of the State Baptist Convention would not embrace creationism enthusiastically, while the apostles of schism on the periphery of religious politics would not let creationism be an issue of secular scientific credibility. Between the former's inertia and the latter's centrifuge, creationism could not achieve a semblance of scientific authority

205

Creationism in North Carolina

in the mainstream culture. To be sure, a generation of children in a hundred or so fundamentalist academies now learns to equate natural science with creationism. Yet those who believe most fervently in creationism have removed themselves from the politics of public education.

For the activists in North Carolina with legitimate scientific credentials who sincerely want the plenary authority of secular science to shed its sanctifying grace on creationist knowledge and belief, the mission of nurturing and advocating scientific creationism is no easy cross to bear. The way they combine the two existential themes of creationism in their own lives is the very stuff of creationism, for here is where creationism becomes a part of one's career, one's family, one's efforts to make a community a better place.

Eleven
A Creationist Study Group

On the second Thursday of the month, for nine or ten months each year, a small group of men gathers at the comfortable suburban home of a science professor from a major North Carolina university. They begin to arrive around 7:15 in the evening and proceed to a paneled den at the back of the house. There they sit in a circle and wait informally, browsing through *National Geographic*, *American Scientist*, or evangelical magazines like *Christianity Today* and *Moody Monthly*. If only about a half-dozen people come on a given evening, they easily make themselves comfortable in the sofa and the armchairs that line the den, but when a dozen or so attend, they squeeze in chairs from the kitchen and sit elbow to elbow. The room has a large brick fireplace in one corner, and a long low stereo cabinet takes up most of the back wall, but the visitors have no interest in entertainment on these occasions. Instead, they come together for serious study of the scientific case for creationism.

Newcomers appear occasionally at these meetings and are welcome when they do, but on most evenings the group consists of the same familiar people. In addition to the host, they include: another science professor at the same university; a computer engineer from IBM, who serves as the group's treasurer; an electronics technician; two lab scientists, one from a pharmaceutical firm, the other from a federal government research facility; and two medical doctors. Others who attend occasionally are college students, electrical engineers, and more lab scientists. Preachers come by rarely.

The host of the group is a man in his mid-forties who

speaks with the slightly nasal twang of the upper Midwest. He chaired the group for four or five years and when not chair is still one of the top officers. His manner is always calm and even-tempered. I never knew him to raise his voice or speak angrily to anyone, or about anyone. Some of the others speak bitterly about evolutionists or recklessly about scientific evidence, but this creationist remains dignified as he makes his way through the creation-evolution controversy, commenting cautiously on difficult issues. For example, he says quite candidly that evolutionists constantly reform their views as more evidence becomes available, adding that creationists should do the same. As a role model with a low-key temperament, he gives the group much stability by gently and indirectly dissuading the more agitated members from saying or doing rash things, especially in public.

The group's ideological leader is the second of the two science professors. The Professor, as I will call him, acts like a stern shepherd steering a fickle flock, constantly reminding the others how important their mission is, and what they must do to achieve their goals, and what they should believe to stay faithful to a certain brand of creationist belief, namely, the views of Henry Morris of the Institute for Creation Research. There is nothing shrill in the Professor's voice. On the contrary, he talks in a flat monotone that makes every matter sound equally serious. Still, he has dead-serious feelings about everything that touches on evolution and creationism, including religion, science, politics, and education. So compelling is his passion for creationism that he finds problems to complain about that the others do not realize are problems.

One evening in spring 1985, the main business of the meeting was done, and the fellows were fidgeting as the meeting was about to end. The Professor got their attention by telling them he had something important to share with them. He showed them an anticreationist book by the anthropologist Ashley Montagu, from which he read a passage about the Arkansas creationism trial of December 1981. According to Montagu, as quoted by the Professor, the evolutionists who went to Little Rock for the trial were having dinner together one evening to celebrate the progress of the trial. After they had had a drink or two, they began singing, and because several had Protestant upbringings, they sang some hymns. At this, the creationists listening to the Professor smiled, but he did not. The Professor told the group that Montagu wrote that the evolutionists sang "Amazing Grace." When they reached the second-verse line about God creating the earth, said the Professor, the evolutionists laughed, because in a way they were singing a creationist hymn. The creationists listening to this tale laughed too, but the Professor frowned angrily. "Hey, guys," he admonished them, "this isn't funny." The other men stopped smiling and looked at him like guilty schoolboys. "This," he lectured, "is the epitome of blasphemy." No one objected that irreverence about a hymn might be less serious than blasphemy. His comment made his fellows somber, and the meeting broke up on that note.

On similar occasions he complained indignantly that books by Stephen Jay Gould were being used in English literature classes, and that the humanist philosopher John Dewey introduced a reading system that made children illiterate. His incessant resentment frightened me, leaving me wondering whether there would be some consequence of pouring all this unhappiness into the creationist cause. And yet I knew him to be extremely kindhearted at other times. He told us one evening that Stephen Jay Gould was suffering from cancer, which saddened him, and he asked us to remember Gould in our prayers.

Over the course of two years, when I saw him often, my feelings about him shifted from fear of his fanatical zeal to respect for his eternal dedication. I saw that his single-mindedness infected the other members of the study group, keeping them focused on the seriousness of their creationist enterprise. Yet this man who is so unrelenting in his ideological leadership and so abrasive in the things he says about creationism's adversaries is also totally modest about his own abilities. Never is he arrogant about his central role in this group, nor does he ever claim personal credit for the consequences of his leadership. Rather, he draws religious piety from these things, always giving God credit for the good things that happen because of his contributions. Perhaps because they see that this strong man is so pious, the other members of the group admire the Professor greatly. They know he has endured job discrimination because of his creationist affiliations, and that he refuses to buckle to harassment or intimidation. They appreciate that he speaks often to Christian groups about creationism, and that he debates evolutionists from time to time. They respect his scientific credentials, and they look up to him for the way he, his wife, and their four children constitute a pious Christian family. Some of these men who have school-age children show how serious they are about conservative Christian values by taking their children out of the public schools and enrolling them in Christian academies. The Professor and his wife have done that, but then they have taken the next step, which is to take their children out of Christian schools and teach them in home education. This is a profound responsibility that, in fundamentalist circles, separates those who are serious about family unity from those who are *very* serious about it.

Also, despite his unwavering disapproval of evolution, and knowing that I still clung to my evolutionist views, he was always helpful to me, answering all my questions and presenting me to other creationists in the friendliest terms imaginable.

Then there is a quiet electronics technician, fifty-five years old, who attends almost every month. He makes hardly any impression at the group's meetings, for his frame is slight, his voice is soft, his health is poor, and his education is modest. He has little expert knowledge to contribute to the doctors, the lab scientists, the engineers, or the science professors in the group. Nevertheless, the Technician, as I will call him, puts his heart and soul into the creationist cause. Before this group was formed, he says, he tried to start a group of his own: "I sent out the word, but it didn't turn out too hot." Although he has no formal training in theology or comparative literature, he spends numerous hours studying his many volumes of the Babylonian Talmud, in translation, plus the *Encyclopedia Judaica*, searching for historical confirmation of various Christian beliefs. He has shown me his correlations linking Talmudic comments about sheep sacrifices with Christian images of Christ as the Pascal Lamb to indicate that the events of Easter week were rooted in Old Testament history. After showing me these notes, he read them to me verbatim and then pressed a copy into my hands, written on yellow foolscap in blue, red, and green ink to highlight three kinds of Talmudic discoveries he made in his studies. Several days after that, he composed and sent me a summary of his views on the religious significance of Jerusalem, which included biblical references, Talmudic references, encyclopedia references, newspaper clippings, maps, and tracts. I do not think that any of his compositions have been published, or that anyone except his wife reads them regularly. Yet this friendly, unassuming man devotes himself to a prodigious task of documenting biblical history, as understood by fundamentalists.

I, too, had a role. If I could have, I would have hidden behind my notebook and my Bible, for what I really wanted was a fly-on-the-wall view of the creationist study group, and nothing more. But that was impossible because these were small, friendly meetings at which the members were extremely gracious. They made me feel welcome among them, often asking for my professional opinions and personal feelings about creationism. I participated modestly by contributing my knowledge of the history of evolutionary and creationist thought, which they liked to hear about but lacked the curiosity to look up. I became by default the group's unofficial historian of creationism. This was a strange thing for me, an evolutionist explaining the history of creationism to creationists, but I was honored and pleased to make myself useful with my esoteric expertise.

On those once-a-month Thursdays, when the Professor senses that it is time for the meeting to commence, he says, "Let's start with a prayer." Conversation ends as the men break eye contact with each other, then each looks down at nothing in particular while listening to the leader's prayer. In an ordinary speaking tone, the Professor tells God that these people have come here tonight to learn more about his creation; he asks God to bless their evening together and enlighten them. If they have a particular concern on their minds that night, such as a group project or a new scientific topic, then he mentions that too in the prayer, asking God to help them with it. Lastly, he states that he says these things in Jesus' name, which is everyone's cue to say "Amen" softly. Now the business of the meeting can begin.

This is the format of a Bible study group. As they begin with an evangelical prayer and later end with one, these folks dedicate their evening together to knowing God better. Many attend weekly Bible study groups on Wednesday nights at their homes or churches. Since this group meets on a Thursday, its mood is often shaped by the spirit of Bible study from the night before. The first time I attended a meeting of the creationist study group, the Professor set the mood by recalling three lessons from Genesis that had been emphasized in his Wednesday night Bible study group—that humankind is created by God, that woman is man's companion, and that divorce is wrong. The next month, just before the opening prayer he referred again to his group's progress in studying Genesis, whereupon he reminded the creationist group that Noah's first act after building an altar to God was to get drunk. He said this should caution them that even righteous men sometimes do things that are wrong: "Some of us become good Christians, but then fall away."

Other times other members contributed reports about creationism from their own Bible study groups. Over many months they kept up an irregular commentary on the meaning of the Hebrew word *yom*, which is the word for "day" in the seven days of creation of Genesis 1. In modern creationist orthodoxy, as handed down by Henry Morris, the days of creation are believed to have been literal twenty-four-hour days. To discourage metaphorical or allegorical interpretations of *yom*, Morris's orthodoxy avers categorically that *yom* always means a day of twenty-four hours in every biblical verse in which it appears. Morris is thus betting the credibility of his chronology on a very narrow line of biblical exegesis. Those members of the local creationist group who attend conservative Bible study groups would say they heard that *yom* always meant a twenty-four-hour day, regardless of its various Old Testament contexts, just as Henry Morris said. Not surprisingly, their conservative Bible study groups encouraged literalist exegesis by doubting contextualist interpretations. Other local creationists who attended less-structured Bible study groups said they heard that *yom* usually meant twenty-four hours, but that sometimes it could be a figure of speech, as in, "in the day of prosperity be joyful"

(Eccles. 7:14) or "the day of the Lord" (Is. 3:12). None of these creationists read Hebrew; in their discussions about *yom* they were simply trading hearsay back and forth. Then one evening a Baptist seminarian told them that *yom* sometimes meant a figurative time, something other than a twenty-four-hour day. Because of his credentials they accepted his opinion. They devalued its implications for Henry Morris's chronology, however, by agreeing that the seven days of creation could still have been twenty-fours hours each even if the day of the Lord or the day of prosperity were not. Later, one of their leaders confused the issue a little further. While the day of the Lord, which is yet to come, *might* be a figure of speech, he said, it could also be a literal twenty-four-hour day; we'll just have to wait and see.

The Bible study format obviously keeps religious issues prominent in the business of the local creationist group, but it also undermines religious orthodoxy by evoking personal opinions about the ways those issues fit into creationist knowledge and belief. Most of the thousands of Bible study groups that punctuate the Protestant population of North Carolina every Wednesday night are informal gatherings in which friends help each other see the meaning of Holy Scripture by contributing individual comments. They encourage do-it-yourself exegesis more than theological conformity, and they permit Christians to bypass the more opaque parts of the Holy Bible by attracting them to the plainer passages, where more Christians can make useful comments. For the creationist group, this means that its members can consider a matter like the meaning of *yom*, but they do not have to resolve it. They discuss it as much as it interests them, after which they shelve it and move on to another topic. Indeed, they might like to resolve issues that are easy to settle by consensus, but they have no reason to face divisive issues, regardless of how critical, that might disrupt their friendly camaraderie.

For example, one evening the topic was to be a lesson about Noah's Flood that would include scientific details about the rain, the animals, the construction of the ark, and so on. The host set a biblical context for the evening by reading Psalm 104, emphasizing verse 6 ("the waters stood above the mountains"). With this he reminded the group that, according to scientific creationism, God had suspended a canopy of water in the "firmament" above the earth on the second day of creation (Gen. 1:6–8); they would see this in scientific terms by examining the climatology and hydrodynamics by which that water was released to become the Flood. But then he anticipated a minor disagreement and defused it by saying that he was reading from the American Standard Version of the Bible, while the language of the King James Bible might not have presented the same picture. In other words, there could be different understandings of what the firmament was, according to which Bible each creationist read. Ironically, this implied that the group would be united more by its scientific conclusions than by its biblical beliefs.

Considering the tight connection between conservative Christian religion and conservative Republican politics in North Carolina, especially in the election year of 1984, I was surprised that this group expressed little interest in politics. One of the founders explained to me, "We're not interested in politics. We're providing information, being informative. If a law was introduced, maybe we could be an expert authority. That's a worthy scientific role." For a while they discussed writing to the local public libraries to have creationist books put on the shelves, but I never knew the group to engage in anything more political than that. Of course they have strong individual affiliations; in the fall of 1984 I grew accustomed to seeing Jesse Helms bumper stickers decorating their cars. But in November the group's meeting occurred two days after the national elections, when conservative Republicans, particularly President Ronald Reagan and Senator Jesse Helms, trounced the Democrats. Yet when the Technician said the opening prayer that evening, he simply asked God to guide our elected officials. He mentioned nothing about liberals, conservatives, Christians, Republicans, or Democrats, let alone the names of any politicians. Neither did any of the other creationists.

Like many small groups, this one has a core of about a half-dozen members who attend almost all meetings and assume most of the mundane responsibilities that must be attended to from month to month. It also has a circle of ten to fifteen members who each attend several times a year, staying active without becoming leaders. Finally, the group has a list of dozens of marginal members and former members, people who used to attend, who used to participate, who still agree with the group about creationism but cannot give it more than a small fraction of their time.

Although the format of their meeting comes from Bible study, the content is technical. These men spend their time together trying to master the empirical evidence for creationism. After the opening prayer, the group settles into an educational program of about an hour's length. Usually they have a packaged program—a film, a tape, or, most often, a slide show produced by the Institute for Creation Research. These packaged programs tell them little that they have not already heard, but they nevertheless study the messages carefully, for they will be expected to bring these programs to their own Bible study groups, their Sunday schools, and their churches. Five or six of the members have spoken often in public on behalf of scientific creationism, including the host and the Professor, so they are comfortable with these materials, which they rearrange and edit to suit themselves, especially the slide shows and the scripts that come with them. The others are less confident, so they stick closely to the scripts. Viewing them in this group's meetings is for them a rehearsal, where they learn to say the right things the right way. There is no opportunity for them to challenge the packaged programs, or even to pose questions to the producers.

Still, the program of the evening sometimes originates within the group. At the first meeting I attended, the topic was the cosmological theory of the inflationary universe as proposed by the physicist Alan Guth. The host led the discussion by summarizing several articles he had read recently in *American Scientist, Science,* and *Scientific American.* Neither he nor anyone else in the room claimed any special expertise in astrophysics. They behaved like any other curious group of intelligent non-physicists trying to understand new developments in physics. Although they certainly wanted to know what Guth's model implied for creationism, no one forced the issue, and they all reserved judgment on it. The Professor reminded the group that popular literature like *Reader's Digest* usually presents scientific ideas as more factual than they really are, while scientific periodicals like *Scientific American* show that many scientific ideas are speculative. No one tried to make Guth's model conform to biblical imagery; no one even mentioned the Bible during that evening's discussion.

Group members want their arguments on behalf of creationism to be scientifically credible, and they worry about this very much. In 1983, the Technician became interested in the theories of an Australian creationist named Barry Setterfield, who said that the speed of light had been much greater in the past, but that it had since decreased, and around 1960 it leveled off at the rate we know today, about 186,000 miles per second. The Technician and others in the study group raised this topic twice, once in late 1983 and again in early 1984. The first time, the host offhandedly cautioned against Setterfield's theories. The second time, he reported that he had asked Henry Morris about it. Morris, he said, had told him that Tom Barnes, the staff physicist at the Institute for Creation Research, was skeptical. After that, the group did not raise the issue again. Perhaps the local creationists should have figured out for themselves that Setterfield's theory is truly a loony idea, without having to hear Morris and Barnes say so. Regardless, they approached this matter cautiously, knowing that they needed expert advice lest their gullibility get the better of them.

They turn to Henry Morris for more than technical expertise: he is their hero, their inspiration. One of the founders of the group, an electronics engineer, traces his commitment to scientific creationism to the time he met Morris in San Diego in 1978. He says it impressed him greatly that Morris, a distinguished fellow engineer, explained creationism and the problems of evolution in terms of their mutual professional standards. Before that, he says, he thought creationism was only a religious idea. His respect for Henry Morris can be measured by an angry comment he made about the judge in the 1981 Arkansas trial. He resented bitterly the way Judge Overton referred to Morris as "Mister" Morris, after having referred to evolutionist experts by the academic title "Doctor." After all, Morris has a Ph.D. in engineering. This man was not so much resentful that a creationist had been slighted relative to the evolutionists, but rather that an engineer had been slighted relative to the scientists. "After reading [Morris's] books," he continued, "I wanted to get this material out to people." When he returned to North Carolina from San Diego, he met with a local professor who had been an engineering student at Virginia Polytechnic Institute during the early 1960s, when Morris was directing one of V.P.I.'s engineering departments. He respected Morris as an engineering professor, as a Christian lay leader, and as a creationist, and he and Morris stayed in contact after Morris left V.P.I. in 1970 to establish the Institute for Creation Research. After this man and the electronics engineer discovered that they had a mutual interest in creationism, they founded the local study group. One of their earliest projects was to bring Henry Morris and another ICR speaker to the Research Triangle area for a two-day seminar in the spring of 1980. During that visit, they also arranged for Morris to preach in two fundamentalist churches, one Baptist and one Presbyterian, that some members of the local creationist group attended.

These creationists also admire Henry Morris for his biblical commentaries on Genesis and on Revelation, lovely volumes bound in smooth leather. They play no role in the group's scientific discussions because their content is traditional biblical exegesis, not technical information, but these volumes are often on display in the home of the host, reminding the local creationists that Morris is a spiritual leader as well as a scientific authority. Thus impressed by his many abilities, they say to each other that it is amazing one man can do so much. This kind of comment leads into folklore about the life of Henry Morris. According to one story I heard several times, Morris is a workaholic who, when fortified by strong coffee, often stays up all night writing. Another tells of his anguish over a son who was dissolute and disrespectful. People could not understand how the son of a leading Christian could be so bad. As the narrative continues, however, Henry Morris perseveres in his creationist ministry and ultimately wins the respect of his son, who then changes his own heart and becomes a reputable Christian.

The occasions of the packaged programs, plus the discussion of the inflationary universe, plus the deference to Henry Morris regarding Setterfield's theory together reveal the attitude of the study group. When dealing with their own expertise and each other's, the people in this group are both as curious and as cautious as any other small gathering of intelligent people trying to learn what science means in our lives. But when receiving knowledge and belief from Henry Morris and the Institute for Creation Research, they defer modestly to Morris's authority, which they never challenge. When dealing with other sources, they proceed cautiously until they get guidance from Morris.

Another dimension of their feelings involves the size of the group and its privacy. When they gather as a half-dozen or so with no strangers present, they can be relaxed about their creationist beliefs and the credibility of evolution; when a dozen or more are together, or when they feel sensitive in the presence of outsiders, they become quite rigid about the beliefs that hold them together. This contrast struck me the first time I met a group of them together, in the fall of 1983. By then I had interviewed about ten of them, each in the privacy of his home or office. Knowing them this way, I had become sensitive to their separate cares and personal doubts. Although each subscribed to the general outlines of creationist belief, I saw, they also differed from each other and their national leaders on some issues. In September I went to a public meeting organized by conservative Christian students at one of the area's universities, where a creationist film was shown. Four members of the creationist group were in the audience, including the host and the Professor. When the student sponsors led the audience into a general discussion of evolution and creationism after the film, I thought the creationists would be as relaxed in public as I had known them to be privately. The host was, and he earned some respect for the cause with his sensitive handling of questions about creationism. But the Professor changed that mood by speaking angrily. He alleged vaguely that evolutionists were suppressing evidence; he replied to evolutionists' comments with a hostile answer, saying, "That's *your* opinion"; when one critic identified himself as a sedimentologist working in coal geology, the Professor implied that geologists who believe in evolution were being subsidized by the oil companies. When the sedimentologist charged that Henry Morris was afraid to

A Creationist Study Group

publish his creationist views in engineering journals, where his peers might pass judgment on them, and that he only wrote "pulp" books for popular consumption, the Professor answered that he knew Morris personally, and that "he would never do anything like that." (It is not for me to say whether Morris is afraid of his fellow engineers, but, as far as I know, he never publishes his views on flood geology and creationist hydrodynamics in professional engineering periodicals.) By the time the meeting was over, I was shocked at how rigid and brittle the Professor had been. And yet the very next month I attended my first private meeting of the local creationist group, when he and his colleagues had such a relaxed discussion of Guth's inflationary universe model.

The contrast between their publicly rigid and privately relaxed moods became apparent again a year later. In October 1984, the group had experienced the encounter with Mister Fossil that I described in the first chapter. That evening was both very public and very humiliating. At the group's next monthly meeting, held as usual in a private home, with six attending, I expected the members to be distressed about Mister Fossil. Their first order of business, in fact, was to discuss what had happened. When the Professor mentioned Mister Fossil, the other creationists rolled their eyes and sighed. The host said that he, too, had been very surprised by what had happened, but that he had not known about Mister Fossil's beliefs previously. The host then said that after last month's event he had asked some people at the Institute for Creation Research about Mister Fossil, and as a result he had a story to tell.

It seems that Henry Morris had invited Mister Fossil to run ICR's Museum of Creation, which at that time consisted of four small rooms at the back of ICR headquarters in El Cajon, California. Morris apparently did not anticipate Mister Fossil's views on biblical floods, and Mister Fossil had not yet seen the ICR museum displays, which featured Noah's Flood prominently, and categorically excluded gap theory, including Mister Fossil's theory of "Lucifer's Flood." The host's story went on to say that Morris and Mister Fossil met first in Morris's office, where they talked about mutual interests and mutual beliefs. It seemed that all was set for Mister Fossil to run the museum.

Creationism in North Carolina

But then Morris showed the museum to Mister Fossil. The visitor sized up the situation immediately and objected to ICR's beliefs about biblical floods. He left quickly and never returned. (This part of the story, incidentally, is remarkably similar to Mister Fossil's own tale about his petulant departure from the Smithsonian many years earlier.)

At this, the local creationists listening to the story laughed. The tension had been broken. True, they had been hoodwinked by Mister Fossil, but so had Henry Morris. With this perspective, their embarrassment was not so bad after all. No one ever mentioned Mister Fossil again.

After recovering from Mister Fossil, they turned the same evening to the main business of the meeting, which was to view a slide show prepared by the Institute for Creation Research that discussed adaptation in the natural world. The show had two things to say about adaptation. First, adaptation is evidence of God's design. When a creature's behavior and anatomy are intimately related to the environmental conditions around it, this can be interpreted to mean that God has planned the natural world very carefully. Second, the slide show emphasized the difference between adaptation and speciation, which it called "*micro*evolution" and "*macro*evolution," respectively. It accepted that adaptation occurs, and that it is observable, as in the case of the English pepper moths, a classic piece of evolutionary evidence wherein one can see a population changing over time in response to environmental changes. By accepting that microevolution is real in the sense that one can observe it directly, it rejected macroevolution by saying that speciation cannot be observed directly.

In the group's discussion following the slide show, one of the creationists surprised me by saying emphatically that the case of the pepper moths proves what the evolutionists allege about adaptation. He went on to say that he believed in *some* evolution, that *some* evolution does occur. Another then chimed in to say that natural selection occurs regularly in nature, just as the evolutionists claim it does. The first man then added that some arguments in favor of evolution are stronger than creationists like to admit, and that creationists should respect that fact. There was silence for a moment, then one of the leaders ended the meeting with the closing prayer.

A Creationist Study Group

This exchange made me wonder whether I was observing creationists or evolutionists. As I thought it over while driving home that night, the only way it made sense to me was to see that the members of the creationist study group could be flexible in their beliefs and relaxed when gathering privately as a small club. It is one thing to judge them by their public appearances, public statements, public personas. Indeed, in public they close ranks and stiffen their spines, admitting no exceptions to their orthodoxy. But in private, surrounded by friends they trust, they are warmer and more tractable.

This I noticed without wanting to experience it, for I intended to be more a detached objective observer than a dear friend of the people I was studying. One evening, however, they warmed me with their friendliness more deeply than I ever thought possible. The main business that night was a creationist film about hominid fossils. Its narrator discussed Piltdown Man and Nebraska Man to make the point that evolutionary assumptions lead to foolish mistakes. He also showed the dental arcades of a young girl and a chimpanzee, so as to suggest that it was easy to overemphasize superficial similarities. The film's narrator then stressed human-chimp differences, for example, their U-shaped and V-shaped mandibles, respectively. He interviewed a British paleontologist who spoke of the difficulties of forensic reconstruction; the film implied that it was unreliable. The movie concluded with the idea that anatomical similarities between species should be interpreted as functional similarities designed by God, not as phyletic links from a common ancestry.

Immediately after the film, the Professor turned to me and said, "Chris, you're an anthropologist. You probably know these fossils better than we do. Maybe you can tell us what weaknesses the film had that we didn't notice because we're creationists." I thought for a moment and replied that this film, like much creationist literature, described differences between hominid fossils in terms of two extreme polarities, labeling them either as obvious apes or as modern humans, with nothing transitional in between. However, I went on, there is a credible continuum of fossil features between the ape-like early australopithecines and the recent Cro-Magnons. I gave the example of KNM-ER

221

1470, an East African skull dated at a little under two million years, which I said was a mix of australopithecine and human features, representing a transition between late australopithecines and early humans. This fossil, I told them, could not be dismissed by labeling it either as pure ape or pure human. I said as tactfully as I could that creationists should not overlook this.

One of the creationists responded that he had heard that Neanderthal skulls fit within the range of modern human variation. Was this true? he asked. "The largest and most rugged modern human skulls," I replied, "are probably Eskimo skulls. Neanderthal skulls are probably more rugged than those." The questioner asked if the Neanderthal skulls weren't only *slightly* more rugged than modern Eskimo skulls. I said, "Well, maybe." At that point another of the creationists, a veterinarian, commented that if Neanderthals were within the modern human range, as creationists say, even if at the end of the range adjacent to Eskimos, creationist scientists ought to be able to find some Neanderthals in the world's population today. He was right, and none of us knew what to say to that.

Next, the Professor asked my opinion about studies of chimpanzee communication. "They show chimps are very clever," I responded, "but they don't prove chimps have human language capacities, or that chimp communication is a prototype of human language." I added that my view is a minority opinion, that most anthropologists are less skeptical than I about this. The veterinarian added that many animals are clever, but that this does not prove evolution. Ironically, he went on to emphasize how intelligent some animals are, telling us of his familiarity with animals, and his concerns about animal welfare in research labs. As I watched him I wondered how far he could elaborate on animal intelligence without inadvertently reaching some kind of evolutionary inference about human-animal similarities and asked myself how he would break off this line of inquiry if it led him to blurt out something favorable to evolution. Suddenly he switched to a rambling tirade about how evolutionists do not want to admit that they are living in sin. (If there was a logical connection from animal intelligence to evolutionists' depravity, I missed it.) Now the discussion had come full circle, back to hearty denunciations of evolution. A doctor seconded the veterinarian's belief about evolutionists' depravity, adding, "I've studied Darwin and the other evolutionists carefully, and I've found that there's nothing in it worth believing. Sure, the pepper moths changed, but that's just genetic variation, not evolution."

Quite a brisk tour of evolution and creationism, I thought to myself. I was relieved that I had been able to answer their questions about fossils honestly without getting myself into a debate against the roomful of creationists. It satisfied me that my comments sparked their discussions without being contentious. After all, I preferred to keep a low profile at these get-togethers.

At this point it was time to end the meeting. I looked toward the Technician, because the Professor usually asked him to lead the closing prayer. He did this, I think, as a gracious way to acknowledge the Technician's dedicated attendance, even though he could not give much scientific substance to the meetings. This time, however, the Professor did not ask the Technician. He turned to me. "Chris," he said, "will you lead us in our closing prayer?"

This stunned me. I thought to say, no, I'm the anthropologist, the observer, the evolutionist, the guy you don't *really* want to lead you in prayer. But I also saw that the Professor was honoring me by asking me. He was telling me that he appreciated my words about fossils and chimps, at least their honesty, if not their substance. This also meant that the creationists' discussions in response to my comments had been good. The evolutionist had not derailed the creationists' meeting. Beyond the particulars of this evening's events, the Professor's invitation meant that he and the group trusted me with the spiritual responsibility, however modest, of leading them in prayer. This tribute I could not decline.

First I stifled my instinct to blurt out a Hail Mary—not the kind of prayer he had invited me to lead. I mentally reviewed the pattern of an evangelical prayer. Over the last several years, I had heard many hundreds, enough to know that they almost always follow a certain four-part paradigm. First, the one praying invokes the person of God the Father, saying "Heavenly Father" or "Our Father in heaven" or something along those lines. Second,

223

it is necessary to say that the people praying are gathered in his name, or in his presence, or in his grace. Third, the one leading the prayer states the purpose or theme of the occasion: "to share fellowship with so-and-so," for example, or "to renew our faith." This is usually the longest part of the prayer. It can include any topic or comment relevant to the occasion, so it varies according to the circumstances and can be any length. This is the place to tell God what is on your mind, knowing that other people are listening carefully to hear it. It is said in a conversational tone, serious but nonetheless personal. To be too reverential, as if bowing before a burning bush, would be wrong. Finally, there is mention of Jesus Christ, which is supposed to remind those present of his sacrifice and the grace we get from it. The most common convention is to say, "We pray this in Jesus' name." At that, everyone mumbles, "Amen."

I took a breath and started in a calm clear voice. "God our Father, as we gather here tonight in your heavenly presence, we're real glad to be able to come together again to study the wonders of your creation, and to share fellowship with each other for this purpose. We're happy that these folks have been able to be here tonight. We don't always understand what you mean in the creation you've given us, and we don't always agree about it. But we're thankful for this wonderful gift you've given us. We say this in Jesus' name."

"Amen."

My prayer drew no comment, so I must have done it right. I was not much of a conversationalist as we moved to the kitchen for coffee and donuts, because I was thrilled that I had been asked to lead the closing prayer.

As I drove home that night I was filled with a double sensation. I was an anthropologist analytically observing human behavior, supposedly, but when the Professor invited me to say the prayer, and when I said it, I was also a creationist. Both at once. This, I thought, is the special sensation that we anthropologists wander into in our work: to be ourselves and one of the people we study, both at once.

I had one last bit of business to consider, which was the ethics of my experience. I did not want to deceive my creationist friends about my work or my beliefs. Again and again, I had

224

told them that I was not a creationist and was not trying to pose as one. When pressed about my personal views on evolution and science, I would say that I am a Catholic, that I get my faith and morals from revelation and inspiration, not from biology or geology or anthropology. To me, evolution is an empirical fact, not a spiritual truth. I know not to search it for God or godliness, I'd say.

On the other hand, I did not exactly wear my faith on my sleeve. Because I wanted to study creationism by observing creationists, I seldom volunteered my own views. The last thing I wanted was to provoke an angry debate with them on evolution and creationism, for that would have jeopardized the happy relations I'd developed with these people. Since the members of the creationist study group saw me at most of the meetings, and since I did not argue against creationism, they might have thought I was with them in their creationist beliefs, or perhaps drifting toward them. I guarded against that misperception by occasionally reminding them what I was doing and why. And yet my dealings with them were too informal for regular official restatements of my beliefs. I tried to hold my behavior to a middle course: not so private that my stance was concealed, but not so formal that my status was irksome. With this on my mind, I feared that saying the prayer would tilt things too much toward the mistaken belief that I was becoming a creationist.

A month later I knew that my fears had been groundless, and that the creationists were not nearly as fickle regarding my identity as I feared. After the evening of my evangelical prayer, they took to introducing me to other creationists by saying, "This is Chris Toumey. He's an evolutionist, but he's our friend."

Twelve
Scientists and Engineers

Another subculture that nourishes and shapes creationism is the occupational world of scientists and engineers. Some of the North Carolina creationists work as scientists, science teachers, engineers, and technicians. One way or another, they have to adjust their professional lives to their commitment to creationism in order to believe that a person can be a good scientist or a good engineer while also being a sincere creationist.

What, then, is science like for a scientist who embraces creationism, and what is engineering about for an engineer who does the same?

To answer questions like these, I interviewed fifty-one individuals who were active in publicly advocating creationism in North Carolina (some of whom were also in the study group). In addition to asking them some standard socioeconomic questions about education and occupation, I also asked how creationism influenced their work, and vice versa. I describe my research methods in the Appendix; here I present those creationists who have occupations in science, science teaching, engineering, or technology.

The educational accomplishments of my fifty-one creationist interviewees include five Ph.D.'s in the natural sciences (two in botany, two in biochemistry, and one in chemistry), one Ph.D. in engineering, and three M.D.'s. Three others were graduate students in Ph.D. programs in the natural sciences (in anatomy, biochemistry, and pharmacology) when I interviewed them between 1982 and 1985, and two more had master's degrees in the natural sciences as their terminal degrees. Another

226

Scientists and Engineers

seven had B.S. degrees in either engineering or the natural sciences (Toumey 1990b:104, 1987:299). Their occupations reflect the same expertise: nine engineers or technicians, six lab scientists, six science teachers or science professors, three medical doctors, and three graduate students in the natural sciences (Toumey 1990b:105, 1987:300).

How do these creationists with scientific credentials exist in scientific environments where evolution is taken for granted as a fact of science? Surprisingly, there is little conflict. Mostly the creationists in scientific occupations work in utilitarian projects far removed from theoretical debates: they breed hybrids, test pharmaceuticals, or control pests, in the cases of three of them. Certainly these research responsibilities require expertise in ecology and genetics, but other topics like the age of the universe or the chart of hominid phylogeny are not exactly daily concerns. Disagreeing with Darwinism has hardly any consequence in this kind of scientific activity, where one can be quite comfortable with artificial selection without having to consider anything about natural selection. Metatheoretical scientific research, where great minds thrash about in disagreements over ontology or epistemology, is one thing, but applied science directed at specific tangible problems is another. Being a creationist may not be an intellectual advantage in utilitarian research, but it is not a disadvantage either.

I asked the scientists whether their creationist views influenced their research, and vice versa. Most said there was no direct connection, but some of the biochemists said the remarkable complexity of the human body convinced them that we could not have evolved through random processes. They recognized that this does not exactly prove creationism. At best, it casts doubt on evolution. (Note that this view defines evolution only in terms of randomness.) Yet there was one man who told of a connection between his creationism and his research. His specialty was the control of crop parasites. This is how he thought about his work:

> Host-parasite relationships are so unique, so intriguing, so complex, that most of the evolutionary explanations are not very satisfying to me as a scientist.

227

Most are void of explanations. I don't publicly express this, but I feel that a lot of research explaining [host-parasite relations] from an evolutionary point of view is a waste of time. It hasn't added anything to our understanding of the relationships. What I'm saying is, from a creationist point of view, if we assume that the relationships were put in there from the beginning, and we get on with studying the interrelationships, time and money could be used for solving those relationships, rather than assuming that they evolved into that relationship. I haven't been able to prove it yet, but I think some assumptions that they did evolve are inhibiting our understanding of the relationships.

I asked for an example. He told me of a case study involving a fungus and a crop. In the early 1970s a type of fungus almost wiped out many varieties of corn. Some botanists felt that if the fungus was adapted specifically to those varieties, then it might be frustrated if the corn crop was altered by crossbreeding the varieties. To prevent corn from self-pollinating, which would prevent hybridization, a genetic alteration known as "Texas male sterility" was introduced into the major varieties, thereby yielding some corn without pollen. When a variety with this alteration was planted in a row alongside a row of unaltered corn from a different variety, then the latter pollinated the former, resulting in a new hybrid. The fungus also attacked the new hybrids, however. In other words, the hybridization had not protected the crop from the fungus. At this point in his narrative, the scientist concluded that his evolutionist colleagues assumed erroneously that the new crop-fungus relationship resulted from a mutational change in the fungus, as if the fungus was adapting to the new hybrids by evolving. But in his judgment, and the judgment of another botanist who was not a creationist, the original form of the fungus had the ability to attack the new hybrids, without benefit of mutation. There were then two hypotheses for the fungus-hybrid relationship: one assumed that the fungus had evolved, and one assumed that its original condition embodied the ability to attack without changing. The former, he thought, was an unwarranted extrapolation motivated by unsubstantiated evolutionary assumptions.

That comment, of course, was a quasi-creationist interpretation of the new fungus-hybrid realationship, as opposed to empirical evidence in support of creationism. Furthermore, the hybridization was hardly designed to be a crucial experiment in which creationist theory directly challenged evolutionary theory. Considering that this man's comment was admittedly more conjectural than empirical, it says little about either evolution or creationism. And yet there is nothing in this account to suggest that the creationist scientist is less likely, or more likely, than his evolutionist colleagues to generate hybrids or eradicate parasites. In isolated projects like this one, scientists are largely insulated from the consequences of either evolutionary or creationist theory. Here it does not matter much whether a scientist is creationist or evolutionist.

This is not to say that conflicts never arise between creationists and evolutionists. In fact, an ugly case of job discrimination has plagued the man I quoted above. This is a serious charge, and I am not conveying a careless accusation from a man with a chip on his shoulder. An independent account of the discrimination substantiates this man's account. The case is set in a science department in a major university. In the words of my creationist interviewee, it went like this:

> Most everyone in the department knows that I'm a creationist. The department head once tried to stop me being involved in outside activities related to it, but another full professor told him he was infringing on my academic freedom. . . . When I was teaching the introductory course, when I first became a creationist, I had told the students that some scientists didn't support the evolutionary hypothesis, that they had an alternative model. I was called in before the full professors, and a statement was made, didn't I know how science works, that most scientists believe in evolution, therefore it must be true. Since then I've been relieved of major responsibility in that course.

As this case developed, the creationist discovered that his salary was considerably lower than his peers' salaries. Although he was a tenured associate professor, he was earning less that some of the department's assistant professors. He attributed this to the actions of the department head, who he said had considerable power to manipulate annual raises.

Another man who worked in the same department told a similar story. He had once been an active creationist, which was well known in the department, but he came to differ with other creationists about certain substantive issues. The first creationist was one of the most committed members of the local creationist group, but the latter man was quite apostate, by creationist standards. Because of this significant difference, I consider them independent sources, such that the second offers credible substantiation for the first. Here is the second man's account:

> I was called in by the department head as a result of one of my [creationist] presentations, which was brought to the attention of the department head. He said it was not the thing to do, it was unprofessional, and I would be watched closely. And I think the full professors got together with the department head to discuss it. And a couple of friends in the department said this affiliation could have had negative influence on my professional development. [Unlike the department head,] they were not hostile; it was a friendly warning.

The department subsequently got a different chair, though for reasons unrelated to those two cases. The first man I quoted told me that the new chair was also anticreationist, but that he had been fair in redressing the problem of salary discrimination and had restored his right to teach the introductory course.

Another of the local creationists, who worked in a biology lab of a major government facility in the Research Triangle Park, reported that his supervisor was intensely hostile to creationism. This creationist had a Ph.D. in a scientific field and was especially articulate on scientific issues, so that the local creationist group wanted him to be their president, which would make him their public spokesperson. He declined, however, citing his fear that his boss would harass him if it were known that he was a creationist. Instead, he agreed to be the vice-president, which would allow him to assume some leadership but remain discreet about his creationist affiliations. I cannot confirm that the supervisor would have harassed this scientist, but his fear of harassment was very real.

The engineers who are active creationists are an interesting group. In her early studies of the modern creationist movement, Dorothy Nelkin noticed that a large proportion of the leaders and activists were engineers, among them, aerospace engineers in Southern California (Nelkin 1976:211, 1977:72–73). She explained the engineers' involvement in creationism in terms of their profession's social and moral concerns:

> Many textbook watchers [including creationists] are engineers employed in the aerospace industry, people who have personally experienced the discrepancy between technological expansion and the ability to deal with the social and economic problems induced by rapid change. They are particularly distressed with the uncertainties and disruptions of modern society, and they associate a "decline in moral and religious values" with the dominance of scientific and secular perspectives. (Nelkin 1982:167)

Two features of the creationist engineers' image of science, according to Nelkin, are that they think of themselves as pragmatic inductivists, whereas evolutionists are armchair theorists, and that they make order and design the central motif of God's creation.

> The creationists claim that applied scientists [that is, engineers and technicians] are interested in creationism because "they have their feet on the ground and are heavily committed to test out theories." Most biologists, they feel, are too "brainwashed" with evolution theory to think flexibly about the evidence. They also argue that people in technical professions,

231

working in highly structured and ordered contexts, are inclined to think in terms of order and design. (Nelkin 1982:73)

A statement that neatly combines engineering, conservative morality, creationism, and fundamentalist Christianity comes from Rev. Jerry Falwell, who once planned to be an engineer:

You could say that science is simplistic because it's exact. I was studying mechanical engineering before I even became a Christian. . . . You come to exact, simplistic answers if you follow the proper equations and the proper processes. . . . Theology, to me, is an exact science. God is God. The Bible is the inspired, . . . [inerrant] word of God. And if everybody accepts the same theses and the same equations, they will arrive at the same answer. (Falwell 1981)

Within my sample of fifty-one interviewees, nine are electrical engineers or electronics technicians, and three are former engineers. They live mostly in the Research Triangle area, which has a dense concentration of high-technology enterprises having strong demands for computer design, computer programming, and other computer services. Five work at IBM.

Their educational backgrounds fall into two classes. Seven entered electronics by way of the armed forces—radar school, for example—or on-the-job training at large firms like IBM and Westinghouse. The other five began their careers with college degrees in engineering or physics. Their work is spread over several areas of engineering: design, programming, manufacturing, testing, sales, planning, administration, teaching, and consulting.

One man ties his engineering work to his creationist beliefs this way:

I was an electronics teacher and technical supervisor for fifteen years in the military. . . . I was a maintenance supervisor of a radar installation. I was responsible for training maintenance technicians, from

232

introduction to basic electronics through maintenance techniques. My last assignment was as maintenance auditor at a space-tracking station. . . . The more I understand electron theory, the more I become persuaded of an intelligence behind it. I was an electronics instructor before I was a believer. I was converted while I was in electronics . . . it reinforced my belief in an intelligent, ordered universe.

Likewise, he relates the second law of thermodynamics to the evolution-creation controversy:

[The second law] means that if you leave things without feedback or attendance, their condition won't improve. They'll deteriorate. As a former electronics technician, [I know that] an electronic oscillator needs a feedback circuit. To sustain oscillations, you need to keep boosting oscillation. To maintain order in a system, you need to keep making adjustments. Now in the creation-evolution controversy, order wouldn't come out of disorder by accident.

Another interviewee has a B.S. in electrical engineering, and his career includes aerospace work in U.S. Air Force missile testing, design work at IBM, and, currently, consulting for energy management. When I asked him about engineering and creationism, he said about the second law of thermodynamics:

There's the classic thermodynamic method and the statistical method. The statistical method conflicts with evolution, it sees entropy as uncertainty. . . . In the DNA chain, you have to preserve the transfer of information. DNA is extremely stable. That's one of its properties. [There is] the impossibility of anything increasing in order if left to itself; evolutionists are glossing over it. These are global laws.

A third person with an electronics background echoed those sentiments: "The second law of thermodynamics means

233

that everything is going from order to disorder. Change isn't coming from a lower level to a higher level. It always goes from higher to lower. The evolutionists, their theory has to say that in the animal world, things go lower to higher, contradicting the second law of thermodynamics."

These creationist engineers spoke repeatedly of order and design in their work, and also of order and design in God's creation, revealing how profoundly important this theme is in their thinking. Consider this statement:

> I learned [in my electronic training with IBM] that there is a certain order, a certain logic, in almost everything around me. I was being trained to work on equipment, and there were certain procedures. Everything had an order and a structure. The same is true in all of nature, in all forms of life. When I look at a piece of equipment as complicated as an IBM typewriter, and the human effort to design it and build it and operate it, it leaves me in awe of what it took to create the things in nature all around us.

At this point, three themes are converging: order in engineering, order in creation, and moral order in society. How is this theme translated from engineering via creationism, or from creationism via engineering, into conservative views on social morality? What are the pathways of logic by which the creationist engineers' three kinds of thinking about order become united into a single, three-in-one understanding of order?

I propose the following model: The social world that God has given us is, like his natural world, an efficient engine created by an intelligent designer for a moral purpose. Like all other engines, however, human society is imperfect. It has a certain irreducible bug built into it, namely, sinfulness, which is to human nature what entropy is to a mechanical device. Furthermore, the division of labor in God's plan excuses him from monitoring or regulating this entropy. That is the responsibility of God's moral maintenance engineers, that is, conservative Christians, especially those in the United States today. In the eyes of the creationist engineers, maintaining moral order against

234

the threat of wickedness is a religious duty equivalent to a serious professional responsibility. Wickedness, in turn, is equivalent to randomness, disorder, and uncertainty.

When the creationist engineers define and denounce evolution in terms of their own professional metaphors, evolution emerges as a silly conjecture that ignores information theory or violates the laws of thermodynamics. Furthermore, their fear of moral randomness implies that a godly society is vulnerable more to internal moral decay than to external enemies like Satan or communism. The subtle bugs and glitches of wickedness and sinfulness—the entropy of human nature—are inherent in the moral system, making the maintenance activities of moral conservatism that much more urgent.

How can entropy be equated with sinfulness? Seeing sin as entropy, and entropy as sin, is a logical consequence of thinking in terms of premillennialism, one of the theological pillars of fundamentalist Christianity in the United States. Premillennialism sees human history as a long, unhappy record of increasing sinfulness since the original sin in the Garden of Eden. Though it anticipates the ultimate triumph of good over evil in a violent apocalypse in the near future, it sees the human past as a progressively depraved and wicked time. As God's children fell from grace into sinfulness, so a closed system falls from initial order into increasing disorder, namely, entropy.

One of the more vexing problems of the modern creation-evolution dispute is to understand why creationists consistently state the second law of thermodynamics in absolute terms; that is, they deny the critical difference between open and closed systems. The answer is that their statements on the second law are not really statements on the second law; instead, they are statements of premillennialist belief, translated into the terminology of thermodynamics. In premillennialism, there are no exceptions to the depraved history of humanity, for God's creation is a closed system. In translation, there are no exceptions to the closed systems that the second law defines. To a noncreationist, it is appalling that theology determines thermodynamic theory, but to a creationist, it is splendid that science and religion dovetail like this.

Projecting their occupational perspective onto their vision

of God and society gives the creationist engineers a two-level model. There is some arrogance in supposing that God thinks and acts like an engineer, that the origins of the natural world and human society must follow blueprints like those for building a mechanical device. But even though they project this much of their own profession onto God, they certainly do not reverse the projection to think of themselves as godlike. Instead, they assume for themselves a more humble stance, a conflation of religious piety with professional roles, in which their responsibility is to monitor the moral level of society and to maintain it against the entropy of sinfulness.

It is true that many in the rank and file of the creationist movement are acutely naive in scientific matters, and that their collective naiveté protects creationism from the skepticism of the U.S. scientific community. The creationist engineers of North Carolina, however, possess a technical sophistication that runs much deeper than many opponents of creationism suppose. Knowledgeable activists like these give the creationist movement much credibility. Also, their talk about thermodynamics and entropy gives the movement a rich vocabulary of metaphors for moral order that neatly blends creationist moral theory with the claim that creationism is well grounded in real science.

One might ordinarily expect creationism to be a prickly thing to embrace for someone who makes a living within the scientific community, which is overwhelmingly anticreationist; and one might expect the values of the scientific community somehow to rub off on engineers, too, so that a good engineer would readily renounce creationist belief. The former premise is largely true, but not entirely. Sometimes, especially in conditions of utilitarian scientific work, a good scientist can smooth the points of friction between scientific habits and creationist belief. I am not suggesting that a person is a good scientist *because* he or she is a creationist. Rather, I mean that sometimes questions of evolution versus creationism can be separated from utilitarian work, so that a person can be a good scientist and, *separately*, a sincere creationist. The subculture of science does not automatically discourage creationism.

So both the scientists and the engineers who are creationists can make their professional lives existentially compatible with creationism. The two kinds of creationists make different professional adjustments, with the utilitarian scientists disarticulating creationism from their work and the engineers blending creationism into theirs. Ironically, the engineers contribute more existential substance to scientific creationism than do the scientists.

Thirteen
Social and Ideological Outlines of Creationism in North Carolina

The past five chapters portrayed the circumstances that have shaped creationism in North Carolina: the peculiarities of state politics and public school education; the dynamics of sectarian support; the subculture of a creationist study group; and the occupational subculture of science and engineering. To complete this attempt to understand creationism, this chapter presents the result of those considerations, namely, a portrait of creationist knowledge and belief in North Carolina during the first half of the 1980s, and of the kinds of people who embraced creationism. The form this portrait takes is a series of statistical measures of creationist sentiment and descriptions of variations within creationist belief.

First, a general portrait of the state at that time: it was the tenth-largest state, with a population of about 6.1 million in 1983, most of whom lived in rural areas or small towns. Whereas the nation was 73.7 percent urban, by the definition of the U.S. census, and the South as a whole was 66.9 percent urban, North Carolina was only 48 percent so. Charlotte, its largest city, was forty-seventh largest in the nation, with about a third of a million residents, but it accounted for only 5 percent of the state's population. Four other cities were over a hundred thousand, but none was even half the size of Charlotte. Meanwhile, the farm population was about 30 percent higher than the national average (U.S. Bureau of the Census 1985:12–13, 10, 365).

North Carolina trailed the national averages in education and income. Fully 66 percent of the nation's adult population had completed four years of high school, and 16.2 percent had

238

finished four years of college; for the state, those figures were 54.8 percent and 13.2 percent, respectively. The national average annual pay in 1983 was $17,545, but the state average was $14,676. North Carolina ranked thirty-seventh among the states in personal income per capita in 1984 (ibid.:134, 418, 440).

And yet the state was far from being uniformly rural, poor, or poorly educated. The metropolitan statistical area of Raleigh, Durham, and Chapel Hill housed three major universities (North Carolina State University, Duke University, and the University of North Carolina), which together constituted a mass of scientific expertise that enticed dozens of major corporations to establish research facilities in the Research Triangle Park. Burroughs Wellcome, IBM, General Electric, and the National Institutes of Environmental Health Science, to name but four, brought numerous highly skilled professionals to that part of North Carolina, with the result that its metropolitan statistical area had more Ph.D.'s per capita than any other in the nation. Reflecting the positive regard for science in the area was an excellent high school dedicated to science education, the prestigious North Carolina School of Science and Math.

The high-tech culture was most dense in the Research Triangle area, but some other parts of the state emulated it also. The Interstate 85 corridor between Charlotte and Durham featured a thin string of corporate headquarters, research centers, and light manufacturing facilities. Because of these, each of the state's five largest cities, which were all on the interstate corridor, was surrounded by a wide skirt of sprawling middle-class bedroom suburbs.

The religious climate of the state was overwhelmingly Protestant in both the rural and cosmopolitan areas. In the state's 100 counties, Baptists constituted a majority of all church members in 44 counties, and a plurality of at least 25 percent in 22 more. Another 29 counties were predominantly Methodist, either by majority or plurality (Quinn 1982:insert map).

Naturally, there was conflict between the rural parts of North Carolina and the newer, more cosmopolitan communities. Religious piety and old-time social conventions together formed a provincial cultural climate in the counties where the interstate highways did not reach. These counties were not necessarily hostile to the high-tech culture of the more cosmopolitan areas, but they were not particularly comfortable with it, either. Contrariwise, middle-class newcomers from the North and Midwest—executives, research scientists, and computer programmers—sometimes worried that a primitive caste of bumpkins occupied the hills and farms that surrounded the new suburbs. If people thought of rural North Carolina in terms of Andy Griffith's fictitious town of Mayberry, then the old-timers believed the ordinary residents were as decent and likable as Sheriff Andy, while the newcomers feared they were as dumb as Gomer Pyle.

These cultural differences should not be exaggerated, for they were often quite benign. In the matter of scientific creationism, they actually complemented each other, each supplying a critical component to this body of belief. The small towns and rural areas preserved a climate of moral conservativism, featuring strong social pressures to enforce puritanical customs, along with the idea that the Holy Bible, in its narrowest interpretations, anchors those customs. The concentration of modern technological enterprises contributed a general respect for science and technology, such that values and customs must not appear to be in conflict with scientific authority. It also supplied the cadre of scientists and engineers who expressed conservative Christian sentiments in scientific idioms—thermodynamics, for example. Thus, scientific creationism: a homegrown conservative morality justified in terms of biblical faith but graced by the scientific sanctification of scientific terminology and engineering idioms.

Three Levels of Creationism

There are several levels of creationism. The top level is the national leadership, that is, men like Henry Morris and Duane Gish, well educated and experienced, who work full-time spreading the word. As a result, their message is articulate, urgent, and highly rationalized. The bottom level, by contrast, is the body of public opinion that agrees with the creationist leaders, without their intense investment of either time or concern. This is the simplest measure of creationism: a pollster asks a sample of citizens about creationism and reports its strength as a proportion of the population. This kind of support is largely passive, however, and not particularly articulate—inert public opinion. In between are active followers. They do not formulate the famous principles of creationist thought but rather defer to their national leaders' judgment. Still, these are people who make creationism matter in their homes, in their churches, in their Bible study groups, in their communities' school board meetings, and so on. Thus, three levels: active leaders, activist followers, and passive sympathizers.

Public Opinion on Creationism and Evolution

The Carolina Poll is a statewide opinion survey conducted twice each year by the School of Journalism at the University of North Carolina in Chapel Hill. The poll of October 1983 surveyed public opinion about creationism and evolution by asking a question originally formulated by the Gallup Poll of July 1982. It also included such standard variables as age, race, sex, education, and so on, in addition to questions about contemporary political issues.

In that poll, a majority (53.4 percent) chose the creationist reply to the question about origins ("God created man pretty much in his present form at one time during the last 10,000 years"), while 42.2 percent accepted the theistic evolutionist position (evolution guided by God), and 4.4 percent agreed to the atheistic evolutionist view. By contrast, creationist strength in the national Gallup Poll of July 1982 was 44 percent (Toumey 1990b:96, 1987:276).

To learn which demographic factors correlated closely with sentiments on creationism and evolution, I ran a series of chi-square tests, the conventional statistical measure for data like these. The two most distinct kinds of correlations were religious factors and the cluster of education, occupation, and income. Both church attendance and the distinction between Baptists and non-Baptist Protestants were intimately tied to the creation-evolution question: those who attended church most frequently were strongly creationist, and those who did not were strongly evolutionist. The creationist position accounted for a powerful majority of North Carolina's Baptists, while the non-Baptist Protestants preferred the evolutionary viewpoint, by a modest majority (Toumey 1990b:97–100, 1987:276–280).

Regarding education, those with modest educations provided the bulk of the support for creationism, and those with high levels of education accepted evolution. Similarly, by occupation, the evolutionists were overwhelmingly professionals, executives, and teachers. Considerations of income mirrored education and occupation. Those in the lower income bracket were distinctly more likely to support creationism than evolution; support for creationism declined as income rose; those in the highest income bracket were clearly more likely to agree with evolution than with creationism (Toumey 1990b:97–100, 1987:276–280). The cluster of education, occupation, and income was thus as strong as the factor of religion in differentiating creationist from evolutionary belief.

Could creationism have made any difference as a voting issue? Could creationists have used electoral politics to convert popular support into public policy? In the mid-1980s, after all, the political climate of North Carolina politics had a strong religious flavor. The state's conservative Christians were enormously proud that Senator Jesse Helms was a national symbol of religious conservatism, deeply involved in efforts to institute group prayer in the public schools and to overturn Supreme Court decisions on abortion. During this time, the Republican party of North Carolina became the political home for activists motivated by fundamentalist values and for candidates who campaigned on fundamentalist issues. The Republicans used tactics that sought to mobilize fundamentalists as voters, for example, by registering voters at fundamentalist churches during services that praised Senator Helms and President Reagan for their views on school prayer or abortion. Thus it is logical to expect that creationist and evolutionist sentiment would somehow be aligned with differences between Republicans and Democrats.

The poll results contradict that. Creationism showed hardly a trace of political significance. A preference for either evolution or creationism made no difference as to whether a person was registered to vote. Furthermore, registered Democrats were neither more creationist nor more evolutionist than the state as a whole. They embraced evolution and creationism in the same proportions as the rest of the adult population. (Ironically, registered Republicans were slightly more sympathetic to evolution than to creationism.) The only political matter that reflected any distinction was Senator Helms's ratings: those who approved of him were slightly more agreeable to creationism than the statewide standards, and those who disapproved were more agreeable to evolution. Even this connection, however, was moderately weak (Toumey 1990b:97, 100, 1987:276–280).

There was a very slight gender gap, with men more likely to accept evolution and women to prefer creationism. Race was insignificant in statistical terms. Place of residence was among the strongest correlations, with the rural areas harboring strong creationist attitudes, not surprisingly, while the towns and cities had corresponding evolutionist views (Toumey 1990b:98–100, 1987:276–280).

Lastly, there is some information about whether people care very much about creationism and evolution. Although the people who favored creationism were more likely than evolution's sympathizers to discuss it and more likely to proselytize about it, in fact most people on both sides discussed it seldom or only occasionally. Very few on either side tried frequently to persuade others of their views. Regardless of the sensational attention it received in the popular media, the question of creationism versus evolution was not particularly compelling to most people in North Carolina.

In short, creationism was clearly more popular than evolution in North Carolina in the early 1980s, and more so than in the nation as a whole. Religion and education were the most important factors in this controversy, such that conservative religion was the principal source of sympathy for creationism, and highly educated people accepted evolution. This means that conservative religion would have been on a collision course with higher education if the creation-evolution issue had consumed much attention or emotion in the state. It was, however, a very low-key issue. The political dimensions reflected that: this was not a voting issue. (A similar survey conducted in Florida in 1982 found approximately the same results, especially that support for creationism came mostly from those with modest educations, and that this topic had little force in electoral politics [Handberg 1982]). While more North Carolina creationists cared deeply about this issue than did evolutionists, those who seldom proselytized for either view represented most of the population. North Carolina was not exactly on fire over the issue of origins.

Creationist Activists

To know how creationism touches the lives of those who do care deeply about it, one must investigate the middle level of creationism, that is, those who participate actively in the creationist cause without being famous or full-time leaders. Here are the people who make creationism matter in their children's schools, and who make it a public issue in their communities. This is the human side of creationism, the element that makes it more serious than an occasional newspaper headline and more substantial than an item in a public opinion poll.

I explored this level of creationism by interviewing fifty-one people in North Carolina between the spring of 1982 and the spring of 1985 who were earnestly involved in advocating creationism. (Regarding research methods, see the Appendix.) Some were highly active: they preached in churches and debated at colleges; they organized Christians and lobbied politicians; they mastered creationist literature carefully and attended creationist meetings faithfully. Others were less active, attending meetings only occasionally. Still, they made important contributions, for instance, by writing many of the creationist letters to the editor in the state's newspapers. An additional group was only marginally active. They cared very much about creationism but had few talents or little time to give to the cause. In this continuum of activism, people had to balance their contributions to the movement against the demands of their jobs and their families.

In several imporant ways, the kinds of people who participated actively in advocating creationism were unlike those whose support was measured in the Carolina Poll. Whereas the supporters of creationism in the total adult population generally had modest educations, modest occupations, and modest in-

244

Social and Ideological Outlines

comes, my sample of fifty-one activists revealed a group situated solidly in the middle class in terms of those same three criteria. In short, their degrees were respectable, their occupations admirable, and their incomes enviable.

The educations of the fifty-one activists included nine doctoral degrees, eleven master's degrees, and sixteen bachelor's. Only eleven of the fifty-one had less than a four-year college education. The relation between education and creationism for the activists was thus the opposite of that described by the Carolina Poll. The distribution of occupations reflected the same thing. Clergy and engineer were the two most common occupations, but there were six lab scientists, six professors of science or science teachers in public schools, and four additional teachers. Another three were medical doctors, and three more were graduate students in the life sciences. Median family income was $31,000. In terms of education, occupation, and income, this sample of advocates contradicts the common perception that creationists dwell in the lower class, far from the higher levels of education (Toumey 1990b:104–105, 1987:299–300).

As to race and gender, every interviewee was white, and forty- nine of the fifty-one were male. Like the other statistics, these invert the profile of creationists in the general population, in which race and gender had little to do with creationist and evolutionary sentiment. Although this was far from a true random sample, I am confident of my conclusion that well-educated, white, middle-class males constituted the core of creationist activists in North Carolina, and if my research were to be duplicated in another state, I would expect to find the same to be true.

The activists identified themselves as adamantly conservative in their outlook and largely Republican in their affiliation. While this is no surprise, it is one more sharp distinction between public-opinion creationism and activist creationism (Toumey 1990b:106, 1987:304).

The religious profile of the activists, however, conformed closely to the support measured in the statewide poll. Baptists were strongly represented (in fact, Independent Baptists outnumbered Southern Baptists), and so were Presbyterians (among whom four of the seven belonged to the breakaway fundamentalist church, PCA Presbyterian). Other conservative denominations (for example, Pentecostal, Christian and Missionary Alliance, Church of Christ) and local independent conservative chapels or fellowships accounted for almost all of the other activists. These religious affiliations thus attested to the conservatism of the creationist activists in two ways: these were mostly churches with well-earned right-wing reputations, and they exhibited the pattern of schism so typical of fundamentalist congregations.

For all demographic characteristics except one, great differences separated the activists who cared very much about creationism from their broad but passive base of sympathizers. The only common feature uniting the activists with their sympathizers was conservative Protestantism. As a result, the mission of advocating creationism flourished in the sectarian subculture of conservative Christian churches, where creationist moral theory was much appreciated, but its concern for secular scientific credibility was less than compelling.

Creationist Knowledge and Belief

The next thing to examine is the ideological fabric of local creationist belief: Considering the many options and variations of creationist thought generated at the national level, what do the local creationists know or believe, and what are the sources of their creationism? For most of the creationist activists in North Carolina, Henry Morris and his organization, the Institute for Creation Research, are the only important sources of creationist knowledge and belief. For information, they refer to Morris to lead them through Genesis and geochronology; for inspiration, they turn to Morris himself to steer them past doubt and difficulty. No other authority or influence matters nearly as much.

For the fifty-one interviewees, ICR's monthly mailings constituted their single most prevalent source. Other information from ICR, notably the books by Henry Morris and Duane Gish, provided background and detail that augmented ICR's mailings, while the Creation Research Society and its journal, the *CRS Quarterly*, concurred with Morris in all substantive matters, whether technical or theological. The local creationist study group was another source, but this was a conduit of ICR's views rather than a competitor (Toumey 1990b:110–111, 1987:319).

Other national sources of creationist information, such as the *Bible-Science Newsletter*, the American Scientific Affiliation, or miscellaneous creationist sources, reached only modest numbers of activists, not nearly enough to rival ICR's leadership.

To probe the texture of the local creationists' beliefs, I posed a series of name-recognition questions. The results showed the same thing again. Henry Morris's was the name most widely known, while another important national creationist leader, Nell Segraves of the Creation-Science Research Center, drew only a fourth as many acknowledgments (Toumey 1990a:110–111, 1987:321).

The name of Rev. Tim LaHaye was identified by almost as many people as was Morris's name. (In fact, several interviewees knew LaHaye's name well enough to correct me after I accidentally misspelled it on the questionnaire.) LaHaye's fame might have been a collateral kind of evidence that corroborated Morris's, since LaHaye, more than any other preacher, sponsored and nourished Morris's work in its formative years. LaHaye had his own claim to fame, however, having founded a Christian college and written a series of popular polemical books, including *The Battle for the Mind*, so it may be that creationists knew about Tim LaHaye independently of Morris and creationism.

The rest of the name recognition reflects scant knowledge about evolution and the creation-evolution controversy. I presented the names of three background figures, namely, William Overton (the judge in the 1981 Arkansas trial), E. O. Wilson (the leading theorist and popularizer of sociobiology), and Niles Eldredge (the cofounder, with Stephen Jay Gould, of the theory of punctuated equilibria). Overton's name was obscure. Wilson's was slightly more obscure than Overton's. Eldredge's was hardly known at all.

Granted, Judge Overton's time of fame was a short two months, December 1981 through January 1982. But Wilson has been an important figure in contemporary evolutionary thought. If creationists have good reason to denounce evolution for overlooking the moral dimensions of human life, then the sociobiology that Wilson represents so prominently deserves their attention. Yet only four of these fifty-one knew who he was. I found their ignorance about Niles Eldredge even more peculiar. Creationists are fond of citing the theory of punctuated equilibria, either to claim that evolutionists are deriving catastrophism from creationism, or to allege that evolutionists disagree bitterly among themselves about first principles of chronology and causation. In fact, the creationists I interviewed raised the topic of punctuated equilibria much more often than any evolutionists I knew. When they spoke of this theory, they often mentioned Stephen Jay Gould, for he was indeed the most famous advocate of "punk-eek." They knew about his books, his articles, his television appearances, his *Newsweek* cover story: all these they resented. But if knowing the name and the work of Gould's coauthor for the major articles on punctuated equilibria was any indication of what one knew about the subject, then the creationist activists knew very little about punk-eek beyond some superficial trivia associated with Gould's celebrity status.

Three items in the name recognition were figures from the history of controversies about evolution. Robert FitzRoy, captain of H.M.S. *Beagle*, was, admittedly, another obscure man. But Alfred Russel Wallace was a major figure in nineteenth-century evolutionary thought. Because only seven of my interviewees knew anything about him, I suspect that their knowledge of Darwinism was thin. So too with T. D. Lysenko, Stalin's crackpot geneticist. The Lysenko affair is the classic case in the history of science of letting extrascientific forces dictate scientific activity. One might expect creationists, along with others who care sincerely about how a twentieth-century society formulates its science policies, to know something about the dangers of Lysenkoism. Yet only five interviewees could recall anything specific about Lysenko.

To keep this in perspective, I say again that the creationists I interviewed were generally well educated, many possessing good scientific credentials. Their critique of evolutionary thought was sincere and impassioned. They worried deeply about moral issues, social problems, and scientific standards, all of which they cited to denounce evolution. I did not set them up to make fools of them by posing the names of Niles Eldredge, Alfred Russel Wallace, or T. D. Lysenko to anyone unconcerned about evolution and creationism. But even with their concerns and their credentials, they possessed surprisingly superficial knowledge about the details of the thing they oppose so passionately. Indeed they knew much about the media celebrities of evolution, particularly Stephen Jay Gould, Carl Sagan, and Isaac Asimov, but very little about the prickly nuances of sociobiology, punctuated equilibria, or the history of Darwinism. What they knew about creation and evolution was generally limited to what they learned from Henry Morris and the Institute for Creation Research, plus some other sources that amplified the Morris-ICR line of authority.

Next, what exactly do creationist activists believe about creationism as science, and about the teaching of creationism as part of science education? For this I presented the interviewees with a range of five statements about creationism and evolution, asking them to choose the one that came closest to their own views. I offered an extreme creationist statement, an extreme anticreationist statement, and three intermediate positions. Similarly, I presented five statements about policies for science education. According to the most extreme creationist statement, evolution should be omitted from science courses; at the opposite end of the scale, the strongest anticreationist statement alleged that "creationism has no place in modern education."

In the range of statements about creationism and evolution, twenty-six of the fifty-one interviewees chose the most extreme creationist statement ("Genesis provides a truthful account of human origins; evolution is an unfounded myth"). However, 45 percent (twenty-three interviewees) chose the position that "creation and evolution are both unproven theories." In the context of the modern creation-evolution controversy, this statement was still a creationist position, for it diminished the credibility of evolution while claiming scientific merit for creationism. Nevertheless, it was distinctly less contentious than the first creationist statement, which is to say that a moderate kind of creationism may accompany the more extreme kind. Not all creationists were uniformly nihilistic about evolution (Toumey 1990b:113–114, 1987:331).

A moderate creationist position also emerged for policies on science education. Only 14 percent (seven interviewees) wanted to purge evolution from the science curriculum, while more than 70 percent (thirty-seven interviewees) recommended teaching both evolution and creationism in science courses. Of course, both policies were creationist, diminishing evolution and claiming scientific sanctification for creationism. But coexisting with the extreme creationist position, which was perhaps what the enemies of creationism expected to find, was a relatively moderate position, and the latter outweighed the former. This suggests that the legislation that creationists embraced, namely, balanced-treatment bills mandating equal time for evolution and creationism, need not be interpreted as a hidden agenda to erase evolution. There has been widespread suspicion among anticreationists that the balanced-treatment policies were meant to be impossible to finance and administer, so that their effect would be for school boards to drop evolution instead of adding creationism. There may indeed be some creationists who intend this, but I interpret the interviews here at face value, that is, to mean that most creationist activists sincerely want public school science education to present both creationism and evolution, on equal terms.

The Ragged Left Edge of Creationism

Creationists generally assume that those who have become creationists stay faithful to the cause, both because their views are so much more satisfying than those of the evolutionists, and because the creation-evolution controversy is so important that it cannot tolerate backsliding, fence-sitting, or other character weaknesses. The movement is not nearly as disciplined as its friends and enemies suppose, however. A surprising number of the activists I interviewed had apostate views that drifted toward belief in theistic evolution, or revisionist critiques of creationism, or confused impressions of creationist orthodoxy. Other members simply lost interest.

The national leaders of the creationist movement make hard distinctions between their own views and those of the evolutionists, and they deny any legitimate middle ground between the two. The battlelines of this controversy should be clear, they feel, and so they despise theistic evolutionists for drifting off into the cloudy no-man's-land between evolution and creationism. Here, they believe, lies a reprehensible character weakness. Theistic evolutionists are Christians so concerned about being accepted in the secular world that they refuse to admit what they know to be true—that Christians are, by definition, creationists. Again and again, I heard it said contemptuously that theistic evolution is a cop-out, a compromise, an inconsistency. Creationists give a grudging respect to *atheistic* evolutionists for being consistent. As they understand consistency, evolutionists are atheistic in their philosophy, leftist in their politics, and depraved in their morality. This they expect, and they appreciate having their expectations confirmed. But a *theistic* evolutionist, to the creationists, is a horrible hybrid, cowardly and confused, unfit and unwanted.

A backslider is equally bad. Creationists feel that the merits of their case are so satisfying and self-evident that no one in their right mind could reject creationism after having embraced it. That, to them, would be incomprehensible. Several times I asked creationist leaders how common it was for creationists to abandon the cause. (Each time I phrased this as "backsliding.") They could not cite a single case.

Considering this image of creationists marching along in disciplined ranks, uncontaminated by backsliders or compromisers, I was surprised to learn that their movement included revisionists, dissenters, and a few very confused individuals. Among them were three who disagreed with orthodox creationism in the matter of chronology, two who challenged basic assumptions of scientific creationism, and some theistic evolutionists.

First, the chronology dissenters: Henry Morris insists that God created the universe in six literal twenty-four-hour days, at the beginning of time, which was relatively recently. These three items are often stated in negative terms by saying that scientific creationism rejects day-age theory, gap theory, and old-earth theory, respectively. These are litmus-test issues to Morris and his followers. Consequently, I was surprised to hear one of the most active creationists, a dedicated member of the creationist study group I have mentioned, say, "I think creationism should not be limited to those who believe in a young earth, or seven literal days." A peripheral member of the study group echoed the other's feelings when he told me, "Those of us in the middle of the road don't believe in literal creationism, like that God created the world in six literal days." A third individual, who is not involved in the study group, but who has spoken publicly in favor of the scientific case for creationism, insisted that his view "is different from the stereotyped creationist, for example those who argue about the exact length of the days of creation. . . . My approach is different from creationists or evolutionists." Naturally these people had a right to dissent about chronology, even if it was just to quibble about minor details, but in terms of Morris's orthodoxy, their nonconformity was heresy, equivalent to peddling evolutionist chronologies.

A considerably more serious kind of dissent came from an interviewee who discounted the importance of the creationists' concerns: "The relationship between creation and evolution is not fundamental to the Christian faith. There's a wide divergence of feelings among Christians, even just among conservative Christians. It's not a fundamental doctrine of the Christian faith, like the Resurrection of Jesus Christ. . . . There aren't many [in an evangelical Christian students group] who see it as a life-and-death thing."

Equally serious, a man who formerly embraced scientific creationism later changed his mind: "I'm still hesitant about believing [evolutionary] phylogeny, but no one has shown me that creationism is any more scientific than phylogeny. . . . [The local creationist study group is] an organization that says they're scientific, when they're not. . . . I think it's wrong for creationists to say that they're scientific in public meetings, when many points of evidence were relating just to the Bible. That's fine for Christians, but that's not science." This man remained a conservative Christian. He challenged the scientific quality of creationism, not its moral philosophy. Nevertheless, his change of heart made him a backslider, as creationists use that term.

Then there were a few theistic evolutionists salted in among the creationists. The first was a doctor with a good grasp of scientific facts and issues concerning evolution and creationism. Of all the interviewees, he was most knowledgable about both sides of the controversy. In the name-recognition section of the questionnaire, he recognized the most names, including Robert FitzRoy, A. R. Wallace, and T. D. Lysenko. This man attended the meetings of the creationist study group more often than most others, though not as often as the officers. He also wrote a letter to the editor in the local paper, publicly advocating creationism. In addition, he was active in a local evangelical ministry. All of his credentials as a creationist were impeccable. Yet from the statements about creationism and evolution, he chose the theistic evolutionist statement, that "the act of creation and the process of evolution are both parts of God's plan."

A second theistic evolutionist was an undergraduate at one of the local universities. He, too, had written a letter favoring creationism in a major North Carolina newspaper. Yet when I asked him about his beliefs, he revealed himself to be tentative and confused about the difference between evolution and creationism, and especially about theistic evolution. He understood scientific creationism to be "a guided evolution with a predetermined spot to end up at"—a muddled view of theistic evolution. He also chose the theistic evolutionist position in the range of five choices.

One more theistic evolutionist was a chemist who was marginal to the local study group, occasionally attending meetings but not participating actively in them. He received ICR's monthly mailings and owned some creationist books. From these, he knew the creationist arguments about thermodynamics, dinosaur footprints, and so on. He shared some creationist views on evolution, particularly the complaint that evolutionists are elitist, arrogant, and authoritarian. "I'm appalled," he said, "that evolutionists have pieced together some facts and some theory, and then present it as fact. I agree with the creationists, that evolution has religious elements." By all these traits, he seemed to be very much a creationist. Nevertheless, he carefully separated himself from the creationists by stating that "people who interpret the Bible literally need creationism," then pointing out that he needed neither biblical literalism nor creationism. He went on to discuss creationism as a manifestation of fundamentalism, mentioning that fundamentalism bothered him. For example, he recognized that creationists consider the creation-evolution controversy to be a so-called pro-family issue, but he did not. "It's associated with those issues," he stated, "because conservative fundamentalists hold these views. But I don't see it as intrinsically entwined [with pro-family issues]." In the item on the questionnaire about policies for science education, he chose the statement recommending that, in the public schools, "science courses should include evolution, and creationism belongs in courses on comparative religion or comparative literature." This, however, is worse than the status quo, according to formal creationist ideology, for it denies scientific credibility to creationism and it demotes the Holy Bible to the status of secular literature.

Having depicted some departures from creationist orthodoxy, I should mention an area in which ICR's ideological discipline operated very effectively to keep creationists' thinking in line. This was the realm of fringe topics in pseudoscience, like UFOs and ancient astronauts. Henry Morris and his associates strongly discourage those kinds of beliefs. As I understand it, their views come more from theological concerns than from respect for scientific consensus: they fear that those topics are competitors to Christian belief. Some fundamentalists believe that those fringe topics are credible but then discourage them on the grounds that Satan is behind them. I mentioned in an earlier chapter that the Bible-Science Association is quite credulous regarding fringe topics, but the Institute for Creation Research rules them out.

I knew only three creationists who exhibited any attachment to such topics. One said that he and his family were studying "theomatics," which he described as a body of proofs that the Bible is mathematically perfect. For example, when references to Satan come in combinations of six (sixth word within certain verses; sixth verse within certain chapters; sixth chapter within certain books, and so on) or when references to Jesus come in combinations of eight, then this is seen to be proof that the Bible has a perfect mathematical structure. Another person spoke of the Bermuda Triangle and other ideas of Charles Berlitz, telling me, "I'm an avid reader of Charles Berlitz, who investigated mysteries of the past, like why the pyramids were built. But the teachers [in high school] gave the traditional view, with no debate, no discussion. For example, they never mentioned that there was a nuclear war in India four thousand years ago, and I'd like to know why it happened. But they just gave a straight, noncontroversial view."

A third man, whom I met at the local creationist study group, spoke at length about his personal interest in pyramid power. After extolling its mysterious strength, he checked himself by saying that pyramid power might have a satanic origin, so he would have to be careful about it.

Still, these are only three people out of many. Regarding fringe topics, the other local creationists accepted ICR's strictures loyally.

Orthodoxy and Contradictions

When it is recognized that there are three levels of creationism, namely, the national leadership, the local activists, and the broad base of passive support, then a full appreciation of the creationist movement in modern America includes all three. In addition, relations and contradictions among the three levels are important to note and understand.

In North Carolina during the first half of the 1980s, the general population, as sampled by the Carolina Poll of Fall 1983, exhibited distinctly more support for a creationist account of origins than for either kind of evolutionary account. Conservative Protestantism was the clearest and strongest predictor of creationist sentiment, not surprisingly, while a high level of education typically resulted in a preference for evolution. Thus the creation-evolution controversy could be reduced, in demographic terms, to a conflict between conservative religion and higher education.

The creationist activists of North Carolina, however, were

255

mostly well-educated, white males of the middle class, with considerable credentials and occupations in science, engineering, and technology. They differed in most respects from the people who constituted their general base of support. It is not hard to see why it mattered so intensely to them that creationism should be accepted as a legitimate body of scientific knowledge. After all, the problem of combining conservative Christian morality with the modern values that celebrate science and technology was hardly an abstraction to them. It was the very stuff of their jobs, their family lives, and the lessons they wanted their children to learn in school.

And yet the single commonality that tied them to their potential supporters was not science or feelings about science but rather religion, particularly the conservative Protestant feeling that where you find immorality, you find evolution.

As to the substantive content of their beliefs, most had a narrow range of knowledge and belief that began and ended with whatever was handed down from Henry Morris and the Institute for Creation Research. With this exception, most of the active creationists were quite uninformed about the broad historical and philosophical outlines of the creation-evolution controversy.

Among those who subscribed to the Morris-ICR hegemony, many exhibited an intense enmity to evolution, granting it no merit at all, but a considerable number felt that creationism and evolution deserved approximately equal credibility. When feelings about creation and evolution were converted into policies for public school science education, the moderate creationist position was especially strong. Most of the activists were quite willing to allow evolution a place in the science curriculum alongside scientific creationism.

Finally, there was much unorthodoxy among the creationist activists. Morris's orthodoxy defined theistic evolution, day-age theory, and old-earth theory as heresies, yet there was an underside where this orthodoxy and discipline evaporated, where theistic evolutionists and other kinds of heretics occupied many of the nooks and crannies of the local creationist movement. None threatened to subvert or depose Morris, but creationist belief was a lot more interesting and varied than most creationists admitted or realized.

Fourteen
Reflections

The people who intend to be God's own scientists want science to make sense morally, at least as they understand science and morality. Certainly there is room in our nation for lusty debates about whether Secular Humanism and the fundamentalist opposition to it are coherent moral systems. But the premise that science ought not to be an immoral institution is good common sense for a twentieth-century society. Science is and must be shaped by cultural values of right and wrong, and it is intimately tied into the fabric of our social life. And so I suggested in my introduction that we have an anthropology of science that asks what moral meanings are invested in science, what symbols make those meanings tangible, and which kinds of people believe in a given system of meanings and symbols.

As to moral meanings, creationism has two overriding themes, namely: an unquenched hostility to the idea of evolution, based on the belief that evolution is intimately involved with immorality (even if cause and effect are unclear); and a quasi-religious awe of science (which creationists share with so many other Americans), so that creationism will be made more credible by the sanctification that supposedly flows from the plenary authority of science.

In an abstract sense, those two themes might be timeless, but in their particulars they are highly specific to U.S. culture in the late twentieth century. The moral theory lapsed into dormancy after the humiliation of the Scopes trial of 1925 but was reanimated in the 1950s and 1960s by the U.S. Supreme Court decisions on school prayer and Bible devotionals, by

257

abrupt changes in sexual behavior, by crises of authority in the years of the Vietnam War, and by other events. Even if it is entirely far fetched to blame these on evolutionary thought, one should see that, in the eyes of a conservative Protestant, those decades were a great battle between organized good and organized evil, with evil eventually labeled "Secular Humanism." I do not suggest that the creation-evolution controversy was as compelling to most fundamentalists as were the issues of abortion or school prayer, but it had to be somehow accounted for and included in that theory of moral struggle, usually by diagnosing the idea of evolution as an extension of Secular Humanism.

Creationist stances on the plenary authority of science have also been time dependent. Earlier in this century it seemed that the Protestant model of science had been entirely eclipsed by the secular model, but more recently it has become clear that the secular understanding of science is not particularly well grounded in national values. Instead, the public understanding of science has been trivialized and fragmented—the trivial model, as I call it—so that the plenary authority of science can be borrowed by co-opting the superficial symbols of science. For creationists, this development raised the possibility that the Protestant model of science could be reestablished after a lapse of a hundred years, if creationism could expropriate some of those symbols. Then there would be, once again, a three-part unity of conservative Protestant morality, science, and public understanding of science.

Thus the leaders of American creationism have great respect for the power of scientific authority in the modern world. They know that, with a secular religion called science coexisting alongside the nation's Judeo-Christian creeds, nothing—not even the Holy Bible—can be credible to the public if it is thought to contradict the mysterious authority of modern science. Almost everything bought and sold in the United States today must make some claim to the approval of the people who wear white lab coats and boast advanced degrees, who dwell in shiny labs decorated with test tubes, cathodes, and computers, and who sanctify our merchandise, whether tangible products or religious values, by blessing it in technological jargon. No doubt

Reflections

many of these claims to scientific sanctification are utter nonsense concocted in the extravagant minds of advertising executives, but our public discourse nevertheless demands them, for ours is a society that takes science too seriously, without caring much what science actually is or how scientists really think. Creationism accommodates this cultural reality by presenting itself as a scientific product.

At the same time, claiming science for creationism has an unintended side effect. By alleging that science substantiates the Bible, modern creationism inadvertently makes the Bible answer to secular scientific standards. Earlier I gave the example of Henry Morris's off-the-cuff exegesis on Gen. 19:26, wherein he explained that Lot's wife became a pillar of salt through chemical replacement. In another of these idioms, Noah's Flood is a discourse on hydrology, his ark-building project is a problem in naval architecture, and his gathering of the animals is a lesson in wildlife management. These explanations tell creationism's constituency that the Bible is more believable when it has a semblance of being scientific, but they also displace the spiritual meanings that make the Bible a moral document. If Noah endured a flood for scientific reasons of climatology, then God's intentions recede to the background, and if the story of Lot's wife was a naturalistic phenomenon, then God's role in that event is irrelevant. This is not a well-recognized feature of creationist thought, for creationists often overlook the intellectual consequences of their views, and anticreationists focus their fears on threats to scientific integrity; for example, when a creationist categorically rejects human evolution on the grounds that Gen. 1:27 says, "God created man in His own image." In a broader historical and philosophical perspective, however, the custom of presenting creationism as a scientific product is a profound concession to secular standards. Scientific creationism has opened the Bible's back door to scientific, quasi-scientific, and pseudoscientific idioms that sneak in and undermine its spiritual messages.

What, then, of the symbols of science? Getting creationism into science education has been the central goal of the modern creationist movement, for one of the most powerful symbols of scientific authority is to have one's knowledge and beliefs

259

taught as science in the public schools. If creationism is to be measured as a political force that changes things in U.S. society by imposing its own understanding of science and morality, then its successes and failures in the public school science curriculum are the best indicators of its influence. It is one thing for conservative Christians who are scientists to persuade their non-scientist coreligionists that science substantiates biblical truth, but this is little more than fine-tuning fundamentalist belief. It is something different, something very daring, to convince the rest of the country.

Presumably there could be several ways to go about this, including publishing creationist articles in science journals, winning over the science departments in colleges and universities, presenting exhibits in public museums, or persuading the National Science Foundation and the rest of the government science bureaucracy to recognize creationism's merits. Here creationism has been either inactive or unsuccessful. It has dedicated enormous effort to public school science education, however, with numerous local successes and failures throughout the nation. The public schools are less beholden to scientific professionalism than are the other institutions and so offer opportunities to make end runs around the anticreationist consensus of the scientific community. Also, public education is a common cultural ground for most Americans. If scientific creationism can establish a foothold in high school biology, let us say, then many millions of people will consider it favorably. Furthermore, creationists think of elementary and high school education as a moral platform on which good and evil struggle for the soul of the next generation. The simple assumption is that youngsters believe the things they are exposed to in school, so that presenting creationism is equivalent to converting kids, and not presenting it is tantamount to abandoning them to evil ideas like evolution.

Then there is the question raised by the sociology of knowledge: What kinds of people embrace which forms of creationist knowledge and belief? Various sectarian factions have different moral interpretations of evolution. Several national creationist organizations promulgate different stances on scientific authority. The national leadership is somewhat different from the local activists, who are in turn quite different from their passive sympathizers. All this variation is reflected in the numerous permutations of creationist thought. In short, modern U.S. creationism is a sprawling, multifaceted, not very neat matrix of belief, and the reason is that many different kinds of fundamentalists and evangelicals involve themselves in creationist moral theory and its stance on scientific authority.

According to Clifford Geertz, religion presents us with two kinds of understandings about the world around us: on the one hand, "a synopsis of cosmic order" that explains how things are and why that is so; on the other, an inspiration of what to do about the way things are. Metaphysics versus ethics, so to speak, also known as models *of* reality and models *for* reality, respectively (Geertz 1973:87–125).

Modern U.S. creationism has both kinds of models, as well as a problem getting from one to the other. Creationism's model of reality, in its most general form, is the not so contentious idea that God has given us a framework of morality within which our lives make sense. Indeed, this is a central assumption of Judeo-Christian thought. But in the United States in the twentieth-century, it seems to conservative Christians that this reality has not been presented realistically enough, for if it is stated just a little too abstractly, then it becomes vulnerable to multiple interpretations by liberal Christians, theistic evolutionists, and other supposedly soft headed people. To make that reality less hazy, more like a solid rock, the theological conservatives wrap it in their epistemology called biblical inerrancy, wherein it becomes both lucid and tangible. If Noah traveled in a wooden boat, like one we can touch, over a stormy sea, like one we can feel, eventually to reach a solid patch of ground, like one we can stand on, then his story has a plain authenticity that, presumably, can be extended to the rest of the Bible's contents. Scientific creationism contributes to this assumption of authenticity by fine-tuning it, making Bible stories as tangible as test tubes. This way, the grace of scientific sanctification enhances the scriptural basis of conservative morality.

And so, within this framework of reality erected by

conservative Christians, there is a special role for the creationists, especially those with bona fide scientific credentials. Their job is to erect a scientific barrier between the Holy Bible's simple truths and the modern moral ambiguities that assault them, by showing both that evolution is immoral and that creationism is scientific. This is not in any sense an invitation to discover new realities. Instead, it is a commission to employ new facts, new jargons, and new symbols in defense of the old reality, which, by definition, is just fine the way it is. In its own way, this kind of contribution can be just as personally rewarding as the kinds of breathtaking intellectual activities we attribute to people like Darwin, Einstein, or Heisenberg. When one scientific creationist turns paleontology upside down by claiming that humans lived alongside dinosaurs, and another reorganizes astrophysics on the principle that the speed of light has been declining sharply since the beginning of time, they give themselves two kinds of satisfaction. First, that scientific evidence has been put right, as it were, so that it supports the inerrantist chronology in which the Creation occurred only about ten thousand years ago. Second, that mainstream paleontologists and astrophysicists—the people who scoff at biblical inerrancy—have been exposed as fickle people hopelessly blinded by evolutionary beliefs about dinosaurs living long ago and uniformitarian ideas about the speed of light. Both of these satisfactions, whether specious or not, affirm the conservatives' model of reality.

Their model *for* reality arises from within the model *of* reality: if the Bible is an authentic history of God's creation, and if science corroborates it as such, then scientific creationists have a Christian obligation to make this known to the world. "Is a candle brought to be put under a bushel, or under a bed, and not set on a candlestick? For there is nothing hid, which shall not be manifested; neither was any thing kept secret, but that it should come abroad" (Mark 4:21–22). Thus, a model *for* reality, that is, a commission to convince the nation that scientific creationism deserves credibility. But how to convince? In general there are two approaches, each treating the content of creationism differently. First is the strict scientific approach, which presents creationism as an issue on its own, abstracted

Reflections

from other moral concerns, asking the public to accept it on its scientific and educational merits. The religious considerations remain in the background as the quotes from Popper, the paradigms of Kuhn, the secular credentials, and the references to thermodynamics or chemical replacement come forward. Often this strategy works, but it has failed in several critical instances, as when leading scientists say its scientific merits are hard to find, and federal judges say its religious qualities are hard to miss.

The second approach is to give creationism a contextual importance as a moral issue alongside abortion, school prayer, pornography, and other matters that concern conservative Christians, usually by framing evolution as an extension of Secular Humanism. Then the preachers come to the fore, mentioning science in passing but dwelling on theology. The good thing about this approach is that the issue of creationism is comfortably surrounded by a host of supposedly related topics from which it borrows moral credibility and, occasionally, political muscle. Creationism gets the benefit of the doubt when in this kind of company. Unfortunately, however, it quickly becomes a low priority in the hierarchy of conservative morality, for abortion and school prayer are always recognized, even by most creationists, to be more important. True, creationism had its day in the sun in 1981 and 1982, but since that time it has receded to the lesser ranks.

A more serious problem with the explicitly moral approach is that creationism is driven back into being a model *of* reality, that is, a closed discourse among people who already accept that the Bible is an authentic history of God's creation. When Jerry Falwell, Jimmy Swaggart, and other white-hot preachers present creationism so forcefully, the many millions who disdain those preachers also disdain creationism's sectarian sources. Evoking images of the Scopes trial, the nonfundamentalist Christians shun creationism as a form of fundamentalist foolishness. Thus the model *for* reality ruins itself by being too blatantly sectarian, thereby leaving the creationists and their preachers to persuade their congregations of what they already believe. This is not such an insignificant thing, for millions appreciate hearing the creationist message again and again. Yet the format of preaching to the converted has its own frustrating problems, which I described in chapter 10. When the creationist message reaches moderately conservative congregations, its general moral message is welcomed, but its fundamentalist elaborations, notably the inerrantist epistemology, are not, and neither are its political implications, such as the plan to force it into public school policy. In other words, moderately conservative Christians know how to keep fundamentalist Christians at arm's length. Here is the biggest frustration the creationists face, for their hopes are aroused when they encounter other Christians but then are dashed as those people cautiously reject the more contentious parts of the message. Things go much better in the true fundamentalist congregations, but these groups frequently dissipate their own sectarian influence as they beget themselves from schism upon schism. Often they seem to be acting out Mark Twain's tale about the preacher who denounced so many people as irredeemable sinners that the godly remnant seemed so small that it was hardly worth saving at all.

Caught between the two models are the scientific creationists of North Carolina. The model of reality that they prefer to project into their schools, churches, and communities is the relatively simple formula of creationist orthodoxy handed down by Henry Morris and ICR. But in their study group meetings and their public presentations (like Mister Fossil's visit), they find that simple formulas do not fit neatly into life's complications. Two unavoidable complications are the subculture of sectarian politics in North Carolina and the state's policies for public school education. As a result, the models and strategies devised nationally are frustrated locally. Furthermore, the local creationist activists are in danger of being isolated within a realm of fundamentalist churches. Certainly this is the opposite of their intention to be God's own scientists. As scientists and friends of science, they want to break out of that narrow sectarian circuit, but as creationists they are disdained by most of their colleagues in science, by most of the public school science teachers, and even by many Protestants.

In a way, that frustration is poignant, for the creationism of the local activists is much more than a model, a theory, or a strategy. It is the existential stuff of their lives, the glue that binds together all the disparate selves of a self-respecting scientist or engineer, a righteous Christian, a dutiful parent, and a good citizen. Their creationism makes them whole. It is the thing that, to them, makes a person all those things at once, and all for the same reason. It is, for them, a personal way in which science makes sense morally.

Appendix: Research Methods

The documents that made possible my historical passages are identified in the text and the references, and the Carolina Poll of fall 1983 was the source of the quantitative measures of creationism in North Carolina in chapter 13 (for which I am deeply grateful to Jane D. Brown of the UNC School of Journalism). My principal research methods were: (1) a series of structured interviews with fifty-one creationist activists in North Carolina; (2) interviews with national creationist leaders in California and elsewhere; (3) interviews with local politicians, scientists, educators, and religious leaders in North Carolina; (4) participant-observation with the local creationist study group described in chapter 11; and, (5) participant-observation at other creationist venues and religious services in North Carolina. My relations with the activists and the study group deserve some description.

To contact the people who were actively and publicly involved in advocating creationism in North Carolina, I began by collecting letters to the editor in favor of creationism in the state's newspapers, especially from March and April 1981, when creationism was often in the news because of laws passed in the legislatures of Arkansas and Louisiana, and from December 1981 and January 1982, the time of the Arkansas trial. When I was able to find the addresses of the letter writers, I wrote to them or called them, asking to interview them. Almost every one of them agreed. While doing this I met several members of the creationist study group. One of the group's officers had spoken to my anthropology class at the University of North Carolina in 1981; this plus my interviews apparently convinced the group that I was researching creationism in good faith, although they knew I was not a creationist. Later, when I made contact again with that group, I asked for their membership list, and they kindly gave me a copy. With this I contacted more people to interview. Some of my interviewees suggested additional creationists for me to interview. From among the letter writers, the members of the study group, and the secondary references, I interviewed fifty-one individuals between the spring of 1982 and the spring of 1985.

In my contacts with these people, I did everything I could to make my feelings and intentions clear. I told them I was not a creationist, that I was not representing myself as one, and that I disagreed with much of what they believed; however, I continued, it distressed me to see that creationists were sorely misunderstood by journalists and evolutionists. I said that I wanted to overcome the cardboard caricatures of creationists, for example, the perception that creationism today was merely a reprise of the Scopes trial of 1925. In a written statement I offered to interviewees, I said, "The purpose of this research is to explore the social and cultural side of scientific creationism, i.e., to know what kinds of people support the creationist argument, and why they feel the way they do about it. . . . It is a search for an honest and accurate description of the movement, which will replace or correct the usual stereotypes of its members and the simplistic cliches about their beliefs."

Some doubted that an admitted evolutionist could do justice to creationism. Others assumed that, despite my honest intentions, "the evolutionist professors in Chapel Hill" would compel me to write an anticreationist dissertation. Nevertheless, they graciously gave me much personal information about themselves and their creationist sentiments. I told them that the interviews would be confidential, that I would protect their identities so they could say things freely. Many said this protocol did not matter. A frequent comment was, "I say the same things in public that I say in private." This was true. Almost all of my creationist interviewees were serious proselytizers with a strong evangelical ethos. Aside from a few items of personal information, such as family income, they felt that there was nothing in my interviews that needed the protection of anonymity.

Research Methods

The ethos that made them so open also made them articulate interviewees. Most had often proselytized before about creationism, the folly of evolution, the immorality that comes with evolution, and so on. They had their own short set-piece speeches on these topics, which they were glad to present to me during the interviews. My questions often elicited comments and beliefs that they had developed and maintained carefully.

After word got around about what I was doing, the creationist study group invited me to come to its meetings, which were held about ten times a year, usually on the second Thursday of the month. I went to thirteen meetings over a year and half (fall 1983 through spring 1985), attending about as frequently as the group's officers, and more frequently than most other members. I participated modestly by contributing practical information, such as contacts with preachers or other conservative Christians I had met who wanted information from the group. I also became, by default, the group's unofficial historian of creationism.

I also attended a number of creationist meetings, and religious gatherings that touched on creationism, as described in chapter 9 and elsewhere. During that time, especially 1983, 1984, and the first half of 1985, I was probably at more creationist venues in North Carolina than anyone else, including the most active creationists. (This does not reflect poorly on them; it only means that I was working intensely on my dissertation while they were balancing their creationist activities with their responsibilities to their jobs and families.)

I had one mildly unpleasant experience, when an interviewee explained to me that, because I was a Catholic, I would be going to hell. Aside from that, I found the creationist activists and other conservative Christians of North Carolina to be exceedingly friendly, gracious, and more than generous in providing me with the information I requested, including personal information about their own lives. I enjoyed their company then, I recall it fondly now, and I will always appreciate the help they gave me.

I still maintain some contact with the local creationist study group. Although I hardly ever attend their meetings, I have made several presentations to share my findings with them since finishing my fieldwork, and occasionally I get constructive criticism from them about my articles and papers on creationism.

269

References

American Humanist Association 1933. "Humanist Manifesto I." *New Humanist*, May–June 1933.
———. 1973. "Humanist Manifesto II." *Humanist*, September–October 1973.
Armstrong, Herbert W. 1983a. "The Key to Human Survival!" *Plain Truth*, December, 1ff.
———. 1983b. "Founder's Statement." *Ambassador College Catalog, 1983–1985*, 9–19. Pasadena, Calif.: Ambassador College.
———. 1984. "Why All These Churches?" *Plain Truth*, June, 1ff.
Beckford, James A. 1975. *The Trumpet of Prophecy*. New York: Wiley.
Bentley, Michael L. 1984. "Creationism through the Back Door—The Case of Liberty Baptist College." *Science, Technology & Human Values* 9(4): 49–53.
Bethell, Tom. 1977. "Darwin's Mistake." *Christianity Today*, 17 June, 12–15.
Bolles, T. H. n.d. "Humanism: America's Greatest Enemy." White Bear Lake, Minn.: Victory. Tract.
Brand, Leonard. 1979. "Presenting the Case for Creation." In *Our Real Roots*, ed. L. R. Van Dolson, 15–20. Washington, D.C.: Review and Herald.
Brown, Robert H. 1981. "Editorial: Scientific Creationism?" *Origins* 8(2): 57–58.
Bryan, William Jennings. 1922. "William Jennings Bryan on Evolution." *Science*, 3 March, 242–243.
———. 1925a. "Who Shall Control?" In *The Memoirs of William Jennings Bryan*, 526–528. Philadelphia: United Publishers of America.
———. 1925b. "Mr. Bryan's Last Speech." In *The Memoirs of William Jennings Bryan*, 529–556. Philadelphia: United Publishers of America.

References

———. [1921] 1967. "Back To God." In *William Jennings Bryan: Selections*, ed. R. Ginger, 229–231. Indianapolis: Bobbs-Merrill.

Bube, Richard H. 1978. "We Believe in Creation." In *Origins and Change*, ed. D. L. Willis, vii. Elgin, Ill.: American Scientific Affiliation.

Burnham, John C. 1987. *How Superstition Won and Science Lost: Popularizing Science and Health in the United States*. New Brunswick, N.J.: Rutgers University Press.

Chick, Jack T. 1972. *Big Daddy?* Chino, Calif.: Chick Publications. Comic book.

———. 1976. *Primal Man?* Chino, Calif.: Chick Publications. Comic book.

Creation Research Society. 1972. Disclaimer statement. *Creation Research Society Quarterly*, December, inside front cover.

Creation-Science Research Center. 1975. *Creation-Science Report: A Five Year Review*. San Diego.

DeCamp, L. Sprague. 1969. "The End of the Monkey War." *Scientific American*, February, 15–21.

Durkheim, Emile. [1915] 1965. *The Elementary Forms of the Religious Life*. Reprint. New York: Free Press.

Durkheim, Emile, and Marcel Mauss. [1903] 1963. *Primitive Classification*. Reprint. Chicago: University of Chicago Press.

Elliot, Jack R. 1982. "What's Wrong with Science?" *Plain Truth*, June–July, 11ff.

———. 1983. "Evolutionists and Creationists Are at It Again!" *Plain Truth*, February, 7–9.

Eve, Raymond A., and Francis B. Harrold. 1991. *The Creationist Movement in Modern America*. Boston: Hall.

Falwell, Jerry. 1981. "*Penthouse* Interview: Reverend Jerry Falwell." *Penthouse*, March, 59ff.

Gatewood, Willard B. 1965. "Politics and Piety in North Carolina: The Fundamentalist Crusade at High Tide, 1925–1927". *North Carolina Historical Review* 42(2): 275–290.

———. 1966. *Preachers, Pedagogues, and Politicians: The Evolution Controversy in North Carolina, 1920–1927*. Chapel Hill: University of North Carolina Press.

Geertz, Clifford. 1973. *The Interpretation of Cultures*. New York: Basic Books.

———. 1983. *Local Knowledge*. New York: Basic Books.

Ginger, Ray. 1958. *Six Days or Forever?: Tennessee v. John Thomas Scopes*. New York: Oxford University Press.

References

Gish, Duane T., and Donald H. Rohrer, eds. 1978. *Up with Creation!* San Diego: Creation-Life.

Giustiniani, Vito R. 1985. "Homo, Humanus, and the Meanings of 'Humanism.'" *Journal of the History of Ideas* 46(2): 167–195.

Gould, Stephen Jay. 1981. "Evolution As Fact and Theory." *Discover*, May, 34–37.

———. 1982. "Moon, Man, and Otto." *Natural History*, March, 4ff.

Grabiner, Judith V., and Peter D. Miller. 1974. "Effects of the Scopes Trial." *Science*, 6 September, 832–837.

Grave, S. A. 1960. *The Scottish Philosophy of Common Sense*. Westport, Conn.: Greenwood.

Grose, Elaine C., and Ronald D. Simpson. 1982. "Attitudes of Introductory College Biology Students toward Evolution." *Journal of Research in Science Teaching* 19(1): 15–24.

Guth, James L. 1983. "Southern Baptist Clergy: Vanguard of the Christian Right?" In *The New Christian Right*, ed. R. C. Liebman and R. Wuthnow, 117–130. New York: Aldine.

Hadden, Jeffrey K., and Charles E. Swann. 1981. *Prime Time Preachers*. Reading, Mass.: Addison-Wesley.

Hammond, Phillip E. 1984. "The Courts and Secular Humanism." *Society*, May–June, 11–16.

Hand, W. Brevard. 1987. Decision in *Smith v. Mobile*. (Decision includes judgment, findings of fact, and conclusions of law).

Handberg, Roger. 1982. "Creationism, Conservatism, and Ideology: Fringe Issues in American Politics." *Social Science Journal* 21(3): 37–51.

Harrison, Lester H. 1990. "Creationism as a Moral Critique." Paper presented at the meeting of the Southwestern Social Science Association, Fort Worth, Texas.

Hegvold, Sidney M. 1982. "Now Some Want Creation Taught as 'Scientific Theory'!" *Plain Truth*, March, 7–9.

Hollinger, David A. 1989. "Justification by Verification: The Scientific Challenge to the Moral Authority of Christianity in Modern America." In *Religion and Twentieth-Century American Intellectual Life*, ed. M. J. Lacy, 116–135. Cambridge: Cambridge University Press.

Hook, Sidney. 1987. *Out of Step*. New York: Harper & Row.

Hopkins, Joseph. 1974. *The Armstrong Empire*. Grand Rapids, Mich.: Eerdmans.

Hughes, H. Stuart. 1983. "Social Theory in a New Context." In *The Muses Flee Hitler*, ed. J. C. Jackman and C. M. Border, 111–120. Washington, D.C.: Smithsonian.

273

References

Kootsey, J. Mailen. 1976. "Can the Christian Afford Scientific Research?" *Origins* 3(2): 97–100.

Kurtz, Paul. 1985. "Humanism." *Encyclopedia of Unbelief*, vol. 1, ed. Gordon Stein, 328–333. Buffalo: Prometheus.

———. 1989. "The New Age in Perspective." *Skeptical Inquirer* 13 (Summer): 365–367.

LaFollette, Marcel C. 1990. *Making Science Our Own: Public Images of Science, 1910–1955*. Chicago: University of Chicago Press.

LaHaye, Tim. 1980. *The Battle for the Mind*. Old Tappan, N.J.: Revell.

———. 1983. "The Religion of Secular Humanism." In *Public Schools and the First Amendment*, ed. S. M. Elam, 1–13. Bloomington, Ind.: Phi Delta Kappa.

Lang, Walter. 1983. "Twenty Years of the *Bible-Science Newsletter*." *Bible-Science Newsletter*, September, 1ff.

Larson, Edward J. 1985. *Trial and Error*. New York: Oxford University Press.

Lewin, Roger. 1981. A Tale with Many Connections. *Science*, 29 January, 484–487.

Liberty Baptist College. 1985. *Undergraduate Studies 1985–86 Catalogue*. Lynchburg, Va.

Linder, Suzanne. 1963. "William Louis Poteat and the Evolution Controversy." *North Carolina Historical Review* 40: 135–157.

Lindsell, Harold. 1977. "Where Did I Come From? A Question of Origins." *Christianity Today*, 17 June, 16–18.

McCollister, Betty, ed. 1989. *Voices for Evolution*. Berkeley, Calif.: National Center for Science Education.

McGraw, Onalee. 1976. *Secular Humanism and the Schools*. Washington, D.C.: Heritage Foundation. Booklet.

McIver, Thomas. 1989. "Creationism: Intellectual Origins, Cultural Context, and Theoretical Diversity." Ph.D. dissertation, University of California, Los Angeles.

McLoughlin, William G. 1955. *Billy Sunday Was His Real Name*. Chicago: University of Chicago Press.

Mannheim, Karl. [1929] 1936. *Ideology and Utopia*. Reprint. New York: Harcourt, Brace & World.

———. [1924] 1952. "Historicism." In *Essays on the Sociology of Knowledge*, ed. P. Kecskemeti, 84–133. London: Routledge & Kegan Paul.

———. [1931] 1970. "The Sociology of Knowledge." In *The Sociology of Knowledge*, ed. J. E. Curtis and J. W. Petras, 109–130. New York: Praeger.

———. [1926] 1971a. "The Ideological and the Sociological Inter-

References

pretation of Intellectual Phenomena." In *From Karl Mannheim*, ed. K. Wolff, 116–131. New York: Oxford University Press.

———. [1925] 1971b. "The Problem of a Sociology of Knowledge." In *From Karl Mannheim*, ed. K. Wolff, 59–115. New York: Oxford University Press.

Marsden, George M. 1980. *Fundamentalism and American Culture*. Oxford: Oxford University Press.

———. 1989. "Evangelicals and the Scientific Culture." In *Religion and Twentieth-Century American Intellectual Life*, ed. M. J. Lacy, 23–48. Cambridge: Cambridge University Press.

Martin, William C. 1973. "The Plain Truth about the Armstrongs and the Worldwide Church of God." *Harper's*, July, 74–82.

Mayr, Ernst. 1978. "Evolution." *Scientific American*, September, 47–55.

Moore, John A. 1976. "Creationism in California." In *Science and its Public: The Changing Relationship*, ed. G. Holton and W. A. Blanpied, 191–208. Boston: Reidel.

Moore, John N., and Harold S. Slusher, eds. 1970. *Biology: A Search for Order in Complexity*. Grand Rapids, Mich.: Zondervan.

Morris, Henry M. 1963. *The Twilight of Evolution*. Grand Rapids, Mich.: Baker.

———. 1972. "Theistic Evolution." *Creation Research Society Quarterly* 8(4): 271.

———. 1974. *The Troubled Waters of Evolution*. San Diego: Creation-Life.

———. 1977. "The Religion of Evolutionary Humanism in the Public Schools." *Impact*, no. 51, September.

———. 1980. "The Tenets of Creationism." *Impact*, no. 85, July.

———. 1982. "Evolution Is Religion, Not Science." *Impact*, no. 107, May.

———. 1983. "Creation Is the Foundation." *Impact*, no. 126, December.

———. 1984. *History of Modern Creationism*. San Diego: Master Books.

———. 1987. "Evolution and the New Age." *Impact*, no. 165, March.

Morris, Henry M., and Duane T. Gish, eds. 1976. *The Battle for Creation*. San Diego: Creation-Life.

Morris, Henry M., Duane Gish, and G. M. Hillestad, eds. 1974. *Creation: Acts, Facts, and Impacts*. San Diego: Creation-Life.

Morris, Henry M., and Donald H. Rohrer, eds. 1981. *The Decade of Creation*. San Diego: Creation-Life.

———. 1982. *Creation—The Cutting Edge*. San Diego: Creation-Life.

Nelkin, Dorothy. 1976. "Science or Scripture: The Politics of Equal

References

Time." In *Science and Its Public: The Changing Relationship*, ed. G. Holton and W. A. Blanpied, 209–228. Boston: Reidel.

———. 1977. *Science Textbook Controversies and the Politics of Equal Time*. Cambridge, Mass.: M.I.T. Press.

———. 1982. *The Creation Controversy*. New York: Norton.

Neufeld, Berney. 1974. "Towards the Development of a General Theory of Creation." *Origins* 1(1): 6–13.

———. 1975. "Dinosaur Tracks and Giant Men." *Origins* 2(2): 64–76.

Newell, Norman D. 1974. "Evolution under Attack." *Natural History*, April, 32–39.

North Carolina Academy of Science. 1982. "Evolution/Creationism." January. Positon paper approved by mail ballot.

North Carolina Moral Majority. 1981. "Textbook Reviews." Durham. Booklet.

North Carolina Project, People for the American Way. 1983. "Defending the Freedom to Learn: Combatting Censorship in North Carolina's Classrooms." Winston-Salem. Booklet.

North Carolina Science Teachers Association. 1981. "Position on Teaching the Theory of Evolution in N.C. Public Schools." Raleigh. Pamphlet.

North Carolina State Department of Public Instruction. 1930. *Course of Study for the Elementary Schools of North Carolina, 1930*. Publication No. 154. Raleigh: State Superintendent of Public Instruction.

———. 1935. *A Study in Curriculum Problems of the North Carolina Public Schools: Suggestions and Practices, 1935*. Publication Nos. 187–189. Raleigh: State Superintendent of Public Instruction.

———. 1941. *Science for the Elementary School, 1941*. Publication No. 227. Raleigh: State Superintendent of Public Instruction.

———. 1953. *Science for the Elementary School, 1953*. Publication No. 293. Raleigh: State Superintendent of Public Instruction.

———. 1958. *Science: A Resource Guide for Grades 9–12*. Publication No. 320. Raleigh: State Department of Public Instruction.

———. 1985. *Standard Course of Study: Science Component*. Raleigh: State Department of Public Instruction.

Numbers, Ronald L. 1979. "'Sciences of Satanic Origin': Adventist Attitudes toward Evolutionary Biology and Geology." *Spectrum* 9(4): 17–30.

———. 1982. "Creationism in 20th-Century America." *Science*, 5 November, 538–544.

———. 1992. *The Creationists*. New York: Knopf.

276

References

Oberschall, Anthony. 1984. "Politics and Religion: The New Christian Right in North Carolina." *Social Science Newsletter* 69(1): 20–24.

Oberschall, Anthony, and Steve Howell. 1982. "The Old and New Christian Right in North Carolina." Paper presented at the meeting of the Society for the Scientific Study of Religion, Providence, R.I., October.

"Old-Time Gospel Hour." 1981. Transcript of debate between Russell Doolittle and Duane Gish, Liberty Baptist College, Lynchburg, Va., 13 October.

Olsen, Edwin A. 1982. "Hidden Agenda behind the Evolutionist-Creationist Debate." *Christianity Today* 23 April, 26–30.

Olson, Richard. 1975. *Scottish Philosophy and British Physics, 1750–1880*. Princeton, N.J.: Princeton University Press.

Overton, William. 1982. Judgment in *McLean v. Arkansas Board of Education* (LRC 81 322), U.S. District Court, 1982. (Reprinted 1982. "Creationism in Schools: The Decision in McLean versus the Arkansas Board of Education." *Science*, 19 February, 934–43.)

Parramore, Tom. 1976. "Red-Tie Bill and the Wingless Bird." *Meredith* (Meredith College, Raleigh, N.C.), Fall 1976, 15ff.

People for the American Way. 1985. "Attacks on the Freedom to Learn: A 1984–1985 Report." Washington, D.C.

Pollitzer, William S. 1980. "Evolution and Special Creation." *American Journal of Physical Anthropology* 53:329–330.

Presbyterian Quarterly Review. 1858. "The Mosaic Account of Creation, Scientific." *Presbyterian Quarterly Review* 7:129–141.

Profamily Forum. c. 1980. "Is Humanism Molesting Your Child?" Fort Worth, Tex. Pamphlet.

Quinn, Bernard, ed., 1982. *Churches and Church Membership in the United States*, 1980. Atlanta: Glenmary Research Center.

Rafferty, Max. 1969. "Guidelines for Moral Instruction in California Schools." Sacramento: California State Department of Education.

Review and Herald Publishing Association. 1957. *Questions on Doctrine*. Washington, D.C.

———. 1976. *Seventh-day Adventist Encyclopedia*. Rev. ed. Washington, D.C.

Rosenberg, Charles E. 1966. "Science and American Social Thought." In *Science and Society in the United States*, D. D. Van Tassel and M. G. Hall, eds. 135–162. Homewood, Ill.: Dorsey.

———. 1976. *No Other Gods: On Science and American Social Thought*. Baltimore: Johns Hopkins University Press.

Roth, Ariel A. 1974. "Editorial: Why a Publication on Origins?" *Origins* 1(1): 4.

277

References

Rusch, Wilbert H., Sr. 1982. "A Brief Statement on the History and Aims of the CRS." *Creation Research Society Quarterly* 19(2): 181–182.

Sandeen, Ernest R. 1967. "Toward a Historical Interpretation of the Origins of Fundamentalism." *Church History* 36:66–83.

————. 1970. *The Roots of Fundamentalism.* Chicago, Ill.: University of Chicago Press.

Schoepflin, Gary L. 1982. "Perceptions of the Nature of Science and Christian Strategies for a Science of Nature." *Origins* 9(1): 10–27.

Segraves, Kelly. 1984. "The Offense to the Christian in the Public Sector." *Bible-Science Newsletter*, October, 1ff.

Segraves, Nell. 1977. *The Creation Report.* San Diego: Creation-Science Research Center. Pamphlet.

Shulman, Albert M. 1981. *The Religious Heritage of America.* San Diego: Barnes.

Skoog, Gerald. 1983. "The Topic of Evolution in Secondary School Biology Textbooks: 1900–1977." In *Evolution versus Creationism*, ed. J. P. Zetterberg, 65–89. Albuquerque: Oryx.

Skow, John. 1981. "The Genesis of Equal Time." *Science 81*, December, 54ff.

Steep, Clayton. 1981. "Scientists Are in a Quandry about Darwin." *Plain Truth*, February, 20–21.

————. 1982. "Eye-Opening Proof That Evolution Did Not Occur!" *Plain Truth*, August, 27–28.

————. 1983. "For Evolutionists Only." *Plain Truth*, November–December, 19ff.

Stenger, William. 1984. "What Spokesmen for Science Are Not Telling." *Plain Truth*, June, 15–17.

Strickling, James E. 1972. "A Statistical Analysis of Flood Legends." *Creation Research Society Quarterly*, 9(3): 152–155.

Stump, Keith W. 1981. "A Gorilla Speaks Out Against Evolution." *Plain Truth*, December, 7ff.

Toumey, Christopher P. 1987. *The Social Context of Scientific Creationism.* Ph.D. dissertation, University of North Carolina, Chapel Hill.

————. 1990a. "Sectarian Aspects of American Creationism." *International Journal of Moral and Social Studies* 5(2): 116–142.

————. 1990b. "Social Profiles of Anti-Evolutionism in North Carolina in the 1970s." *Journal of the Elisha Mitchell Scientific Society* (N.C. Academy of Science) 106(4): 93–117.

————. 1991. "Modern Creationism and Scientific Authority." *Social Studies of Science* 21(4): 681–699.

278

References

Troyer, James R. 1987. "B. W. Wells, Z. P. Metcalf, and the North Carolina Academy of Science in the Evolution Controversy, 1922–1927." *Journal of the Elisha Mitchell Scientific Society* (N.C. Academy of Science) 102(2): 43–53.

U.S. Bureau of the Census. 1985. *Statistical Abstract of the United States: 1986* Washington, D.C.

Watchtower Bible and Tract Society. 1967. *Did Man Get Here by Evolution or Creation?* Brooklyn.

———. 1981. "Accidents of Evolution or Acts of Creation?" Special edition of *Awake!*, 22 September.

———. 1983a. "Creationism: Is It Scientific?" *Awake!*, 8 March, 12–15.

———. 1983b. "Evolution, Creation, or Creationism: Which Do You Believe?" *Awake!*, 22 March, 12–15.

———. 1985. *Life—How Did It Get Here?* Brooklyn.

Weathers, Mary Ann. 1986. "Challenges to Curriculum Materials and Instruction Techniques." September. Manuscript.

Webb, George E. 1983. "The 'Baconian' Origins of Scientific Creationism." *National Forum* 63(2): 33–35.

———. 1986. "'Facts' or 'Mere Theory'? Continuity Among American Creationists." Paper presented at the annual meeting of the American Anthropological Association, Philadelphia, December.

Whalen, William J. 1962. *Armageddon around the Corner.* New York: John Day.

Whitcomb, John C., Jr., and Henry M. Morris. 1961. *The Genesis Flood.* Philadelphia: Presbyterian & Reformed Publishing.

Whitehead, John W., and John Conlan. 1978. "The Establishment of the Religion of Secular Humanism and Its First Amendment Implications." *Texas Tech Law Review* 10:1–66.

Worldwide Church of God. 1971. *A Theory for the Birds.* Pasadena, Calif. Pamphlet.

Wysong, R. L. 1976. *The Creation-Evolution Controversy.* Midland, Mich.: Inquiry.

Index

Index

282

Index

Index

284

Index

Index

Index

Index

288

Index

About the Author

Christopher P. Toumey, a lecturer in anthropology at the University of North Carolina, spent more than five years talking with creationists, attending their study and prayer groups and lectures, reading their literature, and interviewing their leaders to prepare this portrait of the contemporary creationist movement. His essays on other aspects of the anthropology of religion and the cultural meanings of science have appeared *in Soundings, Science, Technology & Human Values, Creation/Evolution, Social Studies of Science, Books & Religion, Ekistics,* and the *International Journal of Moral and Social Studies.*